Mapping Cross-Border Business Networks in the Asia-Pacific Region

MATTHEW SARGENT

RAND
CORPORATION

For more information on this publication, visit **www.rand.org/t/RRA792-1**.

About RAND

The RAND Corporation is a research organization that develops solutions to public policy challenges to help make communities throughout the world safer and more secure, healthier and more prosperous. RAND is nonprofit, nonpartisan, and committed to the public interest. To learn more about RAND, visit www.rand.org.

Research Integrity

Our mission to help improve policy and decisionmaking through research and analysis is enabled through our core values of quality and objectivity and our unwavering commitment to the highest level of integrity and ethical behavior. To help ensure our research and analysis are rigorous, objective, and nonpartisan, we subject our research publications to a robust and exacting quality-assurance process; avoid both the appearance and reality of financial and other conflicts of interest through staff training, project screening, and a policy of mandatory disclosure; and pursue transparency in our research engagements through our commitment to the open publication of our research findings and recommendations, disclosure of the source of funding of published research, and policies to ensure intellectual independence. For more information, visit www.rand.org/about/principles.

RAND's publications do not necessarily reflect the opinions of its research clients and sponsors.

Published by the RAND Corporation, Santa Monica, Calif.
© 2021 RAND Corporation
RAND® is a registered trademark.

Library of Congress Cataloging-in-Publication Data is available for this publication.

ISBN: 978-1-9774-0742-9

Cover: ktsimage/gettyimages.

Limited Print and Electronic Distribution Rights

This document and trademark(s) contained herein are protected by law. This representation of RAND intellectual property is provided for noncommercial use only. Unauthorized posting of this publication online is prohibited. Permission is given to duplicate this document for personal use only, as long as it is unaltered and complete. Permission is required from RAND to reproduce, or reuse in another form, any of its research documents for commercial use. For information on reprint and linking permissions, please visit www.rand.org/pubs/permissions.

About This Report

The Belt and Road Initiative, the failed Trans Pacific Partnership, the Association of Southeast Asian Nations (ASEAN)-led Regional Comprehensive Economic Partnership, and other high-level frameworks have all attempted to reshape the patterns of business affairs and economic alliances in the Asia-Pacific region through centrally managed planning.

This report explores a complementary perspective, asking how the patterns of preexisting business relationships in the region have developed and how organic patterns of interfirm relationships have evolved to tie regional economics together. It examines these interfirm relationships and maps the business networks that extend throughout the Asia-Pacific region. By doing so, it provides, for the first time, an integrated regional assessment of the network structure of interfirm governance in Asia.

In addition to offering a descriptive account of the structure, the analysis looks at suggestive structural factors that offer insight into the future of regional integration. In particular, it speculates that ethnic Chinese capitalists throughout Southeast Asia may be preferentially selected as partners, thereby offering avenues that would tie these economies more tightly to China's growing sphere of influence. The intent of this study is to provide policymakers with a new lens through which to understand regional integration in the region. It builds on and informs the extensive body of RAND Corporation work on economic competition in the Indo-Pacific, including a series on regional responses to U.S.-China competition in the Indo-Pacific (see, e.g., Lin et al., 2020).

Funding

Funding for this study was provided by the generous contributions of the RAND Center for Asia Pacific Policy (CAPP) Advisory Board.

RAND Center for Asia Pacific Policy

This work was conducted within CAPP. CAPP is part of International Programs at the RAND Corporation. CAPP provides analysis on political, social, economic, and technological developments in and around the Asia-Pacific region. Through research and analysis, CAPP helps public and private decisionmakers solve problems, tackle challenges, and identify ways to make society safer, smarter, and more prosperous. For more information, visit www.rand.org/capp or contact the center director (contact information is provided on the webpage).

Contents

Figures

Tables

Summary

Over the past 20 years, the economies of East and Southeast Asia have more than doubled in size. As they have grown, the activities of the Asian firms have become increasingly interdependent and have integrated with ever-more global supply chains and markets. This market and supply chain integration has led to the formation of enduring partnerships that provide an underlying structure for the regional economy. Rather than operating as atomized firms that interact through arms-length contracts on the open market, firms create enduring partnerships and persistent rivalries that shape the patterns of economic activity both within and between national economies.

This study examines these interfirm relationships and maps the business networks that extend throughout the Asia-Pacific region. By doing so it provides for the first time an integrated regional assessment of the network structure of interfirm governance in Asia. In addition to offering a descriptive account of the structure, the analysis looks at suggestive structural factors that offer insight into the future of regional integration. In particular, it assesses the way that ethnic Chinese capitalists throughout Southeast Asia are preferentially selected as partners, thereby offering avenues that could tie these economies more tightly to China's growing sphere of influence.

The network of board of director interlocks provides a useful entry point to examine the evolving character of the regional economy. While these relationships vary in importance and instrumentality, they provide a useful lens for this study because they are reported for virtually all listed companies in a standardized form, which makes it possible to study both transnational and domestic interlock networks.

Substantial transformations in business practices are evident when comparing director networks of 2006 and 2020. In 2006, one striking feature of this structure was the insularity of these economies. Within the Asian firms sampled, there were only 243 cases in which firms from different countries shared the same director. Korean firms, for example, had no interlocks with foreign firms. By 2020, however, extensive interlock networks had developed throughout Asia. The total number of firms in the sample more than doubled, from 9,111 to 21,552, while the number of cross-border director ties increased almost tenfold, from 243 to 2,624.

One significant feature of business networks in Asia is the difference between the interlock networks of East Asian (China, Taiwan, Japan, Korea, Hong Kong) and Southeast Asian (Thailand, Malaysia, Singapore, Indonesia, Philippines, Vietnam, Laos, Myanmar) nations. Though East Asian economies have three times as many firms listed on global stock exchanges as does the Southeast Asian region, they have only one and a half times as many shared directors.

One cause of the cross-border ties linking these countries is shared linguistic and ethnic ties, which provide both an important basis for cross-border integration in the region and a mechanism that allows Chinese firms to partner with firms that might be less accessible to firms from other East Asian countries. As many scholars have noted, the large population of ethnic Chinese businesspeople in Southeast Asia is a resource that Chinese firms can draw on when entering a new region.

Today, China is actively courting these ethnic Chinese populations. As Xi Jinping observed in a 2016 speech, "There are tens of millions of Chinese overseas compatriots (*haiwai qiaobao*), all of whom are members of one big Chinese family. . . . They are making important contributions to the development of the Chinese nation and . . . fostering the close and friendly cooperation between the Chinese people and people of other countries" (quoted in Suryadinata, 2017, p. 20).

Among business theorists who believe that board interlock is a powerful tool of corporate governance, these networks are thought to "facilitate a community of interest among the elite of the corporate world that supplants the competitive and socially divisive ethos of an earlier stage of capitalism with an ethic of cooperation and a sense of shared values and goals" (Bowman, 1996, p. 21). Even if we adopt the more measured view that director interlocks are "mundane but consequential" (Davis, 1996, p. 154), the disproportionate representation of Chinese firms on boards in the region will likely have long-term positive impacts on Chinese integration into the regional economies, encouraging path-dependent decisions such as the adoption of norms, including Chinese technology standards, that could produce a durable orientation toward China.

However, while the number of cross-border interfirm ties is increasing, particularly between countries with large populations of ethnic Chinese businesspeople, the intensive networking within the national economies of the region will likely exercise a moderating influence by offering some degree of resistance to foreign interventions. Rather than imagining the developing economies of the region as filled with small companies that are at risk of takeover from their larger, expanding neighbors, it is more accurate to think of them as enmeshed in a larger, network ecosystem in which the majority of firms are already aligned within a larger framework. The firms, markets, and even political systems of the economies of the region are already deeply intertwined, and this interconnectedness creates an obstacle for foreign entry. The business groups that dominate many of the economies of the region are natural matrices for the formation of these networks. Because business groups are marked by coordinated action and shared governance, the members of these groups typically place board members on each other's boards to increase information sharing and trust and to foster coordination. In nations with weak formal institutions, these groups are thought to internalize the functions typically provided through public means: The groups provide internal mechanisms for dispute resolution, provide access to financing, and lower the risk of default. Moreover, their economic importance often develops in parallel to strong political ties. In Indonesia, for example, 70 percent of exchange-listed firms are affiliated with one of the business groups, and the head of one the largest, the Bakrie Group, is also the head of a major political party. The result is that there is an extensive and increasingly formalized set of networks in place, which tie the firms of national economies together even as the number of ties between countries increases.

Acknowledgments

I wish to thank Joshua Traub for his technical assistance and Angel O'Mahony and Rafiq Dossani for their expertise and advice.

Additionally, invaluable discussions with James Lincoln, Marleen Dieleman, Han Shih Lee, and several prominent businesspeople and investors working in East and Southeast Asia helped to develop this work. Thoughtful reviews by Michael Pollard and Johan Chu helped clarify and strengthen its arguments.

CHAPTER ONE

Mapping Networks of Influence in East and Southeast Asia

Over the past 20 years, the economies of East and Southeast Asia have more than doubled in size. As the region's economies have grown, their component businesses have integrated with ever-more global supply chains and markets. As firms' activities have internationalized, enduring partnerships and persistent rivalries have developed, and these now underlie the regional economy. As economists and sociologists have long observed, repeated business exchanges give rise to durable relationships built upon trust, obligation, patterns of investments, debts, and shared partnerships (Granovetter, 1985; Williamson, 1985). As a result, rather than operating as atomized firms interacting through arms-length contracts on the open market, firms' networks of relationships shape the patterns of economic activity both within and between national economies.

This study examines the evolving structure of these interfirm networks. While many forms of relationships are difficult to observe, some are formalized in official governance structures, such as shared board of directors' memberships or equity ownership, which render the underlying patterns of relationships visible. These ties give participating firms the potential to control behavior of subordinate firms, monitor business partners, share information, or coordinate action with their partners.[1] These *board interlocks*, as shared director ties are called, and shareholding networks are among the most studied features of the American economy (Chu and Davis, 2016), but their structure and function in Asia are less understood.

The following analysis examines such interfirm relationships that extend throughout the Asia-Pacific region. By doing so, it provides for the first time an integrated regional assessment of the network structure of interfirm governance in Asia. The report is intended to provide a descriptive analysis that highlights the existence and importance of these interfirm networks and is organized around three principle aims. First, it identifies the extent of shared governance ties between Asian firms, both within and between countries. Second, it highlights the regional differences in the extent and structure of these ties. Third, it speculates on the implications of these ties for China's evolving role in the regional economy, both in terms of the structure of the networks connecting China to other countries and the potential role of ethnic Chinese populations in the region in facilitating the formation of these networks.

Governance Networks and International Policy

Governance relationships are important because they confer opportunities for influence and control that can be used to further larger policy aims. Faced with rising protest in Hong Kong in 2019, for example, the Chinese government instructed its state-owned firms to take control of Hong Kong companies by buying additional investments that would enable them to assert control and provide decisionmaking power rather than simply holding minority stakes (Davies and Marlow, 2019) (see Figure 1.1). Air China Limited, for example,

[1] The terms "ties" and "relationships" are used interchangeably in the text.

FIGURE 1.1
Chinese-Held Equity Stakes in Hong Kong Firms

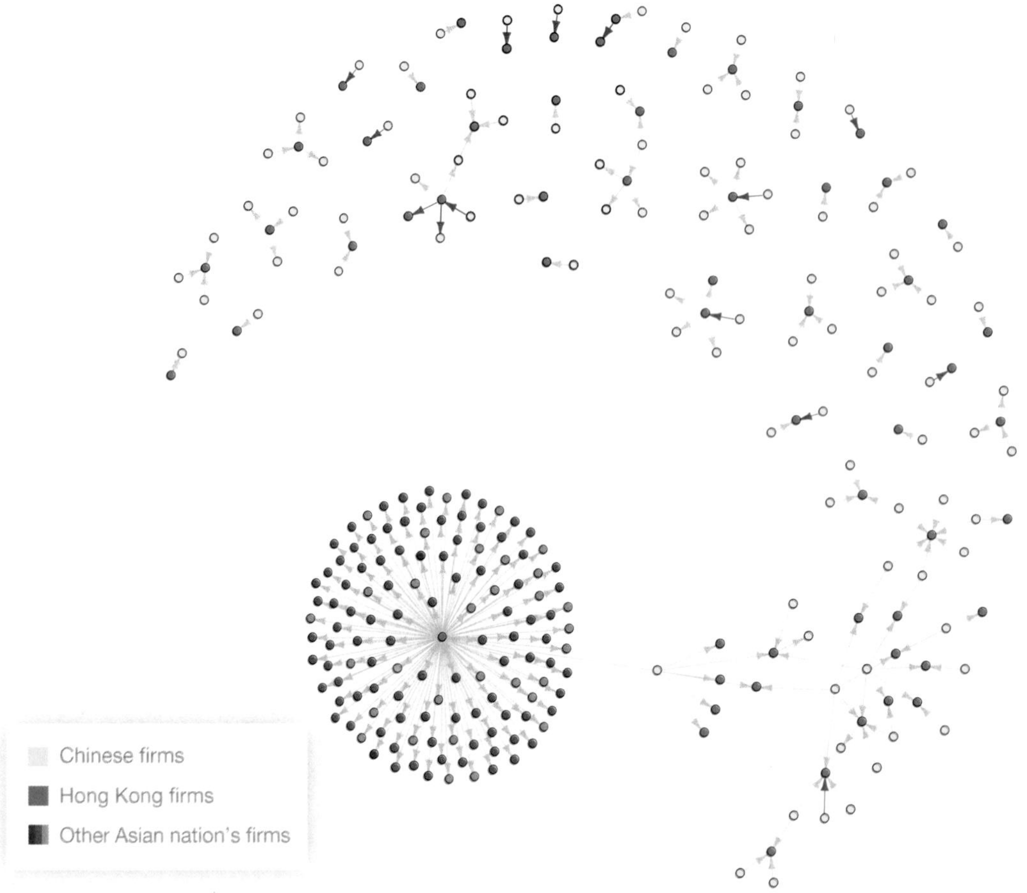

NOTES: This figure depicts Chinese equity stakes in Hong Kong–based firms held by listed firms, private organizations, state-owned enterprises, and individuals as well as the equity stakes held by the Hong Kong–based firms. Gray arrows represent equity stakes of unknown size or of less than 10 percent. Pink arrows indicate equity stakes of 10–50 percent. Red arrows represent stakes of 50 percent or more. The large cluster at the lower left represents the equity investments of the Hong Kong–based AIA Group Limited, which is one of Asia's largest insurance firms. It is connected to the network through a Shanghai-based investment fund, China International Fund Management, which holds AIA stock (the percentage was not reported) as well as stakes in several other Hong Kong firms.

currently owns 30 percent of Hong Kong–based Cathay Pacific Airways, giving the Chinese firm some sway over its partner without providing outright control. By encouraging its existing investors to increase their shareholding, the Chinese government was signaling that it wished to exert direct control, rather than simply a guiding influence, over business partners in Hong Kong. This episode highlights the ways in which business ties can be used to forward policy objectives and underscores the need to better understand the character and structure of these relationships in the Asia-Pacific region.

One striking feature shown in Figure 1.1 is that these governance ties are often not limited to isolated pairs of companies, where a single Chinese investor owns a stake in a single Hong Kong–based firm. Instead, numerous firms are linked together through ties of equity ownership into large networks of affiliated firms. The degree to which these business networks extend across international borders and their potential impacts are not well understood. It is clear, however, that the process of globalization and market integration is not uniform and is instead drawing some national economies closer together than others. Firms' decisions to partner or to collaborate in certain markets are not made at random, and relationships based on factors such as shared language, advantageous policies, mutual distrust, or economic benefit will cause firms to organically integrate and grow closer together.

The existence and structure of these networks are particularly significant, given the number of recent policy initiatives intended to forge durable ties between the economies of the region. In view of the rapid growth of China's economy and its expanding geopolitical influence in particular, many of these policy proposals have been variously designed either to promote or to contain China's role in the regional economy. These include the failed Trans-Pacific Partnership, the United States' Indo-Pacific Economic Vision, the Association of Southeast Asian Nations (ASEAN)-led Regional Comprehensive Economic Partnership, the Comprehensive and Progressive Agreement for Trans-Pacific Partnership, the Asia Pacific Economic Cooperation Bloc, and the Chinese-led Belt and Road Initiative; all have attempted to develop formal mechanisms to forge deeper alliances within the region.

Alongside this focus on formal frameworks, regional integration is simultaneously occurring independently through decentralized channels and is composed of relationships between individual pairs of firms. Firms are the primary economic actors in the national economies of Asia, and as their supply chains and markets have become more international and more intertwined, firms have formalized their partnerships through strategies such as shareholding and shared board members, which allow coordination or coercion across national boundaries. By examining these networks, this report offers a new perspective on the evolving relationships and spheres of influence that underlie the economies of the region.

Asian Business Networks

Interfirm networks are particularly significant within Asia. Historically many of the region's economies have been dominated by business groups or conglomerates that, as one researcher described, stand out as "islands of planned coordination in a sea of market relations" (Richardson, 1972, p. 883). The members of these groups, of which the Chinese *qiye jituan*, the Korean *chaebol*, and the Japanese *keiretsu* are familiar examples, act in concert to support shared group-level strategies. In contrast to the vision of competitive and atomized firm decisionmaking championed by American economic theorists, these multifirm networks are central to the operation of the regional economy.

Most scholarship portrays these business groups as clusters of firms linked to one another in formal and informal ways, including cross-shareholding, interlocking directorates, loan dependence, goods transactions, and social relationships (Chang, 2003; Colpan and Hikino, 2010; Colpan, Hikino, and Lincoln, 2018; Granovetter, 1995, Granovetter, 2005; Lincoln and Sargent, 2018; Yiu et. al. 2007). Groups are thus defined by their connections; firms more densely and closely tied to one another on a variety of network dimensions are visible as densely packed and bounded icebergs in a sea of relations (to use Khanna and Rivkin's [2006] metaphor).

These conglomerates are particularly significant, because they were often controlled by ethnic Chinese families, whose family and shared ethnic ties extended across the region. The concept of this so-called bamboo network was popularized in the 1990s and identified as a driving force behind China's economic growth (Weidenbaum and Hughes, 1996).

As researchers at the time noted, a relatively small group of foreign-born ethnic Chinese business tycoons dominated the business landscape in Asia and preferred to conduct their business dealings within their extended families or deal with other ethnic Chinese businesspeople with whom they shared some connection. The economic influence of this extended network of ethnically Chinese capitalists was unmistakable. In 1999, for example, the businesses controlled by Liem Sioe Liong's Salim Group accounted for 5 percent of Indonesian economic output, while in Hong Kong, Li Ka-shing owned 12 percent of all listed stocks (Borsuk and Chng, 2014; Weidenbaum, 1998). Building on ties of shared language, culture, and ethnicity, these ethnic networks were responsible for substantial cross-border investment flows into China that fueled its early development. In the 20 years after China's economic opening under Deng Xiaoping in 1978, foreign-born ethnic Chinese accounted for roughly 80 percent of all foreign investment in China (Weidenbaum, 1998). The result was a regional economy based on an "ersatz capitalism" that was structured on systematic cronyism, which paired well-positioned families with political patronage networks (Kunio, 1988).

However, the shock of the Asian Economic Crisis of the late-1990s and the internationalization of business over the following decades has transformed the postwar system of family- and relationship-based business connections. In desperate need of external capital due to the economic crisis and with weakening patronage networks resulting from the ensuing political turmoil, many firms shifted from their closely held family structures to adopt a more standardized model in order to better access international credit markets. Foreign financiers required that firms become increasingly "credible" and "transparent" in the eyes of potential creditors, and Asian firms largely adopted management practices and systems of financial control that were prescribed by the gatekeepers of the global financial system (Yeung, 2006). In the words of Peter Woo, successor to one of Hong Kong's largest conglomerates, "There are no friends in finance. The world has changed. They [the old-style Chinese entrepreneurs] need to realize that we are in a world market and need an international culture" (quoted in Yeung, 2006, p. 224).

Despite these changes, the role of family, kinship, and ethnic networks has not disappeared (Luo and Chung, 2005). While firms have adopted the norms of international business, as well as the shareholders, boards of directors, and professional managers that comprise the three central pillars of modern corporate governance (Monks and Minnow, 2012), they have often adopted these structures in form only (Young et al, 2008). Shareholding networks allow formalized control of network companies by an apex firm, just as traditional family-based conglomerates did. Similarly, board seats offer an informal avenue of monitoring and control over subsidiary firms. In other contexts, they may also serve as powerless bureaucratic entities that are filled with the trusted lieutenants of a controlling family. Thus, while many aspects of business operation have modernized to align with global norms, they have not removed the traditional reliance on family and ethnic ties. Rather, these traditional models have, in many cases, coopted these new formal business strictures to create networks that allow business groups to coordinate the actions of member firms both within and across national borders (Colpan, Hikino, and Lincoln, 2018; Yeung, 2006).

As Asian firms have standardized their business practices according to international norms, the structure of their relationships has become increasingly legible. While researchers seeking the identities of the companies affiliated with Indonesian business groups once had to travel to the groups' corporate headquarters to photograph promotional displays that listed member companies, standardized data that have been systematically reported to stock exchanges and collected by international data providers such as the Bureau van Dijk are now available. The network structures defined by patterns of shareholding and shared board of director membership provide a useful, if imperfect, lens through which the relationship structure underlying the regional economy can be observed and analyzed.

A Systematic Approach to Mapping Business Networks

While existing research on Asian business groups has documented the important roles played by these groups in particular national contexts, little work has been done to map the ways these groups extend beyond national borders or to understand the ways that they influence the creation of the cross-border ties that include them. This report uses the standardized data made available as a result of the standardization of business structure to develop a more systematic view of the ways in which governance relationships span national boundaries to create regional webs of influence, coordination, and control.

In decades past, researchers typically used externally produced directories to map business group membership, and it was often difficult to find uniform data on group membership that spanned national boundaries. Instead, research tended to focus on the development, structure, and operation of these groups in particular national contexts, where local historical trajectories and unique social, political, and economic trajectories could be explored (e.g., Colpan and Hikino, 2010; Colpan, Hikino, and Lincoln, 2018; Lincoln and Gerlach, 2004).

Instead of relying on idiosyncratic national data, the organization and impact of these groups can be understood by using the tools of social network analysis (Lincoln and Sargent, 2018). Network analysis is particularly useful in this context because while much of the existing research on business networks struggles with the diverse forms that these conglomerates can take, this approach provides a unified framework and toolset for understanding them. By mapping the structure of the interfirm governance ties that compose these networks, we can study them both within and across national boundaries using a single source of data.

Networks and Governance

Governance ties exist between firms to foster trust, forestall opportunism, and formalize relationships among exchange partners. Efficient economic activity depends on a governance framework that provides for stable, controllable economic exchanges relatively free of opportunism and defection risks. Within business groups, firms are bound together, in part, by formal bureaucratic structures such as equity stakes, board representation, and joint ventures, which enable firms to exercise some degree of formal control and monitoring over their business partners' actions. In addition to these "hard" governance strategies, informal mechanisms offer alternative governance safeguards; reciprocal trade relationships or shared networks of social or business ties may also serve a stabilizing function for interfirm relations, thereby lowering the risk of a partner's default or defection (Granovetter, 1985; Kranton, 1996; Williamson, 1985).

This range of possible channels of influence raises the question of how best to imagine the ways in which central firms might be able to exert control at the peripheries of their groups. Lincoln, Gerlach, and Ahmadjian (1996) propose that the Japanese *keiretsu* are best conceived, less as groups than as "gravitational fields," implying that chains of connections extend from core members to peripheral members. As long as the equity stake in a firm held by the firm one link above it in the chain is large enough to confer control, the influence of the apex firm can be exercised downward through the pyramid with no attenuation of the influence.

However, this preoccupation with clear chains of equity ownership and control can distort our view of the governance and power relationships at work within business groups, because it overlooks the important informal governance controls provided by the myriad other relational ties that exist within groups. As Granovetter (1985, p. 487) argued, business transactions are deeply "embedded in concrete, ongoing systems of social relations" that generate trust and discourage malfeasance. Understood within this framework, interfirm relationships begin with economic exchange and become increasingly "multiplex" over time, as governance relationships, trust, and other ties reinforce and stabilize the exchange relationship and give rise to tightly interwoven clusters of network ties.

While pyramidal groups structured around a single apex holding company are often taken as the standard structural form of business groups (Colpan and Hikino, 2010), many more decentralized network forms also exist. In other groups, horizontal ties among group members enables them to supplement vertical hierarchies with a "network form" of organization, controlled less by vertical hierarchy of authority and more by consensus, coordination, and cooperation among peripheral actors. As Granovetter (1995, p. 96) described, network business groups are "loose coalitions of firms . . . in which no single firm or individual holds a controlling interest in the other firms." This understanding of such a pattern of organization that is less dependent on managerial hierarchies—"webs with no spider" in Lockwood's (1968, p. 503) memorable phrase—overlaps with the considerable literature on agglomeration economies, wherein large dense urban configurations support sizable populations of small and specialized firms that are able to contract externally for many of the functions and services that must be provided internally in firms located in less richly urbanized environments (Piore and Sabel, 1984; Porter, 1996; Saxenian, 1994).

An alternative view, grounded in transaction-cost economics, holds that imbalances in the exchange of goods and services between economic agents lead those agents to craft structures—principally formal acquisition, but also long-term contracts, reciprocal trading, director interlocking, and the like—to offset or control them (Williamson, 1975; Williamson, 1985). From this perspective, visible formal governance structures are an outgrowth of imbalanced economic transactions. In referring to these networks of instrumental relationships, Khanna and Rivkin, (2006, p. 8) suggest that "the bonds put in place to forestall opportunistic behavior may lead to confederations of firms that outsiders then perceive as groups."

These different strategies for coordination and control can all be analyzed through the lens of network analysis (Lincoln and Sargent, 2018). While the topology of the networks alone cannot tell us about the organizing principles or goals of the member companies, it can provide an entry point for analyzing the structural patterns that underly business activity in the region. The conception of groups as densely and completely interconnected regions of a broader interorganizational has considerable intuitive appeal, for it directly speaks to Khanna and Rivkin's (2006) concern that groups be viewed, not in isolation, but rather as topological features of a larger network landscape.

This perspective accords with recent network theory, and research pays close attention to the degree to which interactor ties occur within closely linked and clearly bounded network clusters as opposed to across them. (Watts, 1999). In an economic context, identification of these clusters is important not only as a means of defining group boundaries, but also because several highly influential theories suggest that brokerage or bridging ties that span clusters are superior for information transmission and innovative activity and confer benefits on the actors that maintain these bridges (Burt, 1992; Vedres and Stark, 2010). The location of a firm within these network clusters can thus have important implications for its performance and growth. While some groups may be organized in a hub-and-spoke configuration in which a central actor acts as the focal actor within the network, other organizations might be marked by densely interconnected lower-tier suppliers that allow otherwise peripheral firms to coordinate and innovate without relying on the intermediation of a central network actor.

Analytical Strategy

This research takes firms as the foundational economic actors in the region and studies their evolving relationships with other firms as a means to assess the transnational and regional business groupings that underpin business activities in the region. The research is based on a dataset comprised of all exchange-listed firms in East and Southeast Asia (China, Japan, South Korea, Vietnam, Taiwan, Hong Kong, Malaysia, Singapore, Indonesia, Laos, Philippines, and Myanmar) in 2006 (fourth-quarter data) and 2020 (second-quarter data) with revenue (turnover) of more than $10 million.[2] During the period of our study, economies of the region were growing rapidly, as were the maturing stock exchanges in the countries studied. Thus, while the 2006 dataset includes 50,625 directors in 9,111 firms, our 2020 data included 21,552 firms and 363,929 directors.

As Figure 1.2 illustrates, the dates that we have chosen for this analysis allow us to compare the structure of economic networks that evolved as the economy of the region transformed. Between the 2006 and 2019

[2] *Exchange-listed* refers to publicly traded companies whose shares are traded on a public stock exchange.

FIGURE 1.2
Gross Domestic Product in Select Asian Countries, 2000–2019

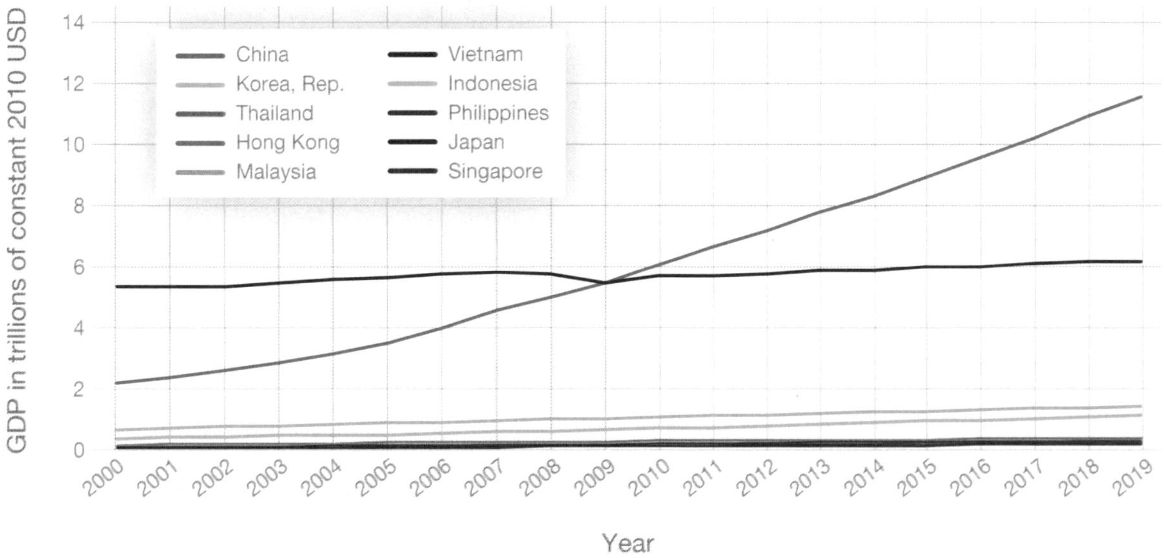

SOURCE: Author's computations of data from World Bank Data, "GDP (Constant 2010 US$)," 2020.

(the latest full year available), the gross domestic product (GDP) of the Asia-Pacific region more than doubled, while China's economy grew by 187 percent.

Shareholder and Director Interlock Networks

Our study focuses on two types of interfirm relationships that represent two of the formal governance methods employed by public companies: the voting rights of shareholders and the board of directors.

The first of these, equity ownership, is straightforward. Companies, banks, other organizations, or individuals can buy stock in another company to both share in its profits and exercise control through shareholder voting. Anything more than a 50-percent equity stake effectively confers control by guaranteeing majority voting rights to a shareholder, though minority stakes also provide the ability to exert influence over firm decisionmaking.

Board interlock, in which one company invites representatives of a second firm to sit on its board of directors, forms a second formalized type of interfirm relationship. Board memberships may be controlled by the firm itself or dictated by outside parties, as when investors and partners demand board seats to protect their investments and interests. Board members are predominantly executives from other firms who typically serve on the boards of one or two other firms (Chu and Davis, 2016; Davis, 1996). These board memberships bring executives into contact with other members with comparable ties and thereby create a broad and complex set of linkages with other firms (Boyd and Hoskisson, 2010). As other researchers have described, this network of relationships "facilitate[s] a community of interest among the elite of the corporate world that supplants the competitive and socially divisive ethos of an earlier stage of capitalism with an ethic of cooperation and a sense of shared values and goals" (Bowman, 1996, p. 21) and represent a "trans-corporate network, overarching all sectors of business" (Useem, 1986, p. 53). More recently, researchers have taken interlocking directorates as a proxy for the existence of a formal link between two companies and have used them to map the structure of business networks (Colli and Vasta, 2018).

In practice, the function of these shared directors varies considerably, and there is significant debate about the true function of board interlocks and how instrumental they really are. Early analyses of these interlocking directorates identified them as mechanisms that enabled collusion between firms and could function as a pathway through which banks could control corporate decisionmaking (Katz, 1998; Pfeffer and Salancik, 1978). More recently, the prevalence of board interlocks in the United States has diminished, and it is no longer clear that they were the major channels of corporate control, but they still perform multiple functions from conferring legitimacy to enabling interfirm monitoring and collusion (Mizruchi, 1996). Moreover, even in their reduced state, they are still understood to be "mundane but consequential" in that they provide a conduit for social influence, the spread of normative behavior, and sharing information (Davis, 1996, p. 154; Granovetter, 1995). Even if they no longer serve as control ties in within the United States, interlock networks provided a communications framework that enabled governance strategies such as the "poison pill" that protected against hostile takeovers to be shared between firms and spawned coordinated adoption of the multidivisional corporate structure (Davis, 1991; Palmer, Jennings, and Zhou, 1993). Others have argued that while corporate boards have the potential to play an active role in corporate governance, they are at other times powerless figureheads and have been characterized as the organizational equivalent of an appendix—an organ that serves no useful purpose (Boyd and Hoskisson, 2010, p. 676).

Within Asia, there is strong evidence that board interlock is still often used to perform important instrumental functions. Within the Japanese *keiretsu*, for example, companies invite representatives of important customers and suppliers to sit on their board, and these ties are an important vehicle of coordinated control (Lincoln and Gerlach, 2004). In China, research a decade ago showed that interlocking directorates resulted from the state appointing key director positions within state-owned enterprises, and these boards were routinely used to share expertise (Lee and Kang, 2010). In all likelihood, these dynamics continue today.

An examination of corporate networks in Singapore concluded that interlocking directorates replaced the traditional informal networks based on kinship and that they "performed the function of co-ordination and control in the marketplace" (Zang, 1999, p. 864). Accordingly, interlocking directorates can be understood as "a modern form of maintaining personal relationships in the Chinese business community" (Zang, 1999, p. 864). While global financial flows and investment decisions are increasingly depersonalized, boards of directors retain their traditional face-to-face interactions and personal relationships. Unlike equity and debt relationships, which can be depersonalized and global, corporate boards typically meet in person several times per year and require direct interaction between the participants. Board composition is therefore more than a bookkeeping entry; it is a record of interpersonal interaction.

While arguing for the significance of board interlock networks, this study does not attempt to differentiate among relationships that are used to exert control and influence, promote coordination, or share information. Instead, the analysis focuses on the structure and topology of the networks with the understanding that they could be used for a variety of purposes.

Methods

Data for this analysis were gathered on all the exchange-listed firms in East and Southeast Asia[3] with annual revenues of more than $10 million USD (current dollars). Data were collected for 2006, 2018, and 2020. Using the Osiris database, published by the Bureau van Dijk, data were gathered on firms, including their name, nationality, industry, as well as size and financial information. The distribution of firm revenues is plotted in Figure 1.3.

[3] The dataset includes firms incorporated in China, Taiwan, Hong Kong, Japan, Korea, Laos, Vietnam, Thailand, Malaysia, Singapore, Indonesia, and the Philippines.

FIGURE 1.3
Distribution of Firm Revenues, 2020

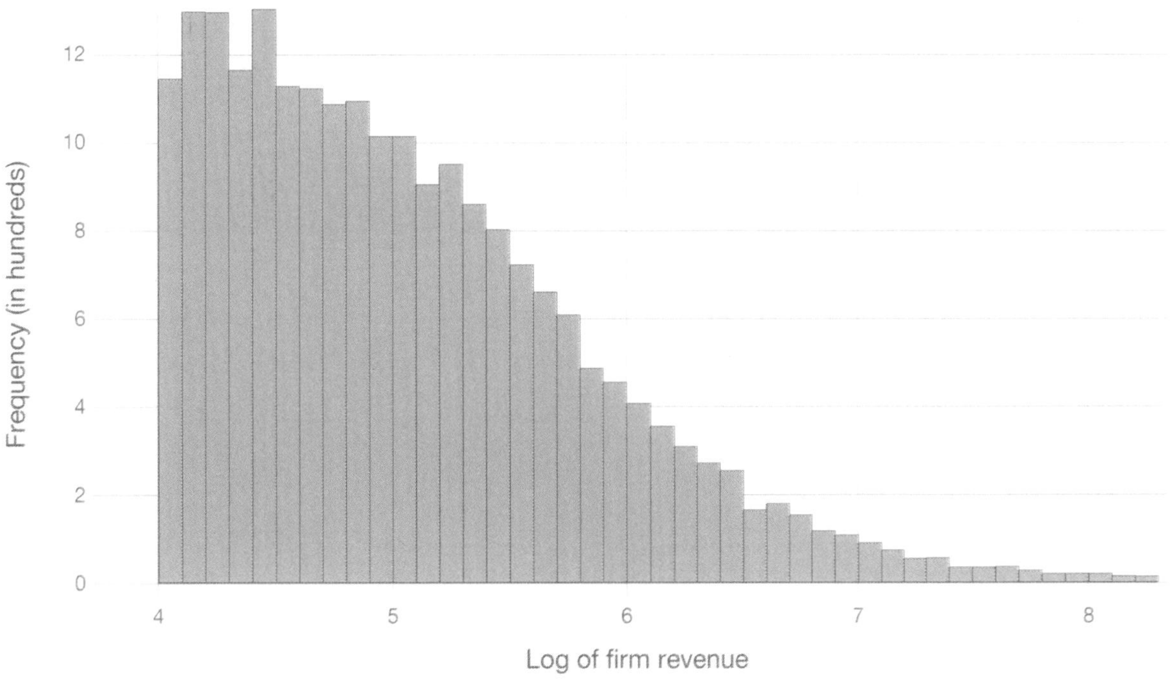

In addition, data on their board of director makeup, including the name, position, and nationality of each of their board members, were collected. Unique identifiers were available for each firm, which made it possible to avoid any problems that would have been encountered from nonstandard naming conventions. Directors were identified by name and nationality; and these were used to create a unique identifier for each first name, last name, and nationality triplet. This name-matching strategy unavoidably creates false-positive matches between directors with the same name—an issue that is particularly prevalent in nations such as Korea and China where a handful of very common family names accounts for the large majority of surnames. In the 2018 and 2020 data, however, there was a unique code for both each firm and each director. The false positives in the 2006 data tend to overstate the number of ties that exist, which makes the relative growth in ties in 2018 and 2020 datasets all the more significant.

There was some regional variation in the number of directors included on corporate boards in different countries. This variation reflects different national norms for board composition as well as different norms and standards for reporting. While all companies reported the positions of board chairpersons, chief executive officers, and outside board members, others also included greater detail on internal board members, typically senior engineers and executives who are employees of the company. The inclusion of these employee board members increased the average number of positions reported but did not affect our overall data since these individuals seldom hold seats in external firms. Additionally, different national rules limit the number of board seats that an individual can hold, and thus limit the configurations of director interlock that develop.

The data on which this report is based were used to create a bipartite network for each year of data using the *igraph* package implemented for the statistical programing language *R* (Csardi and Nepusz, 2006). This approach maps the relationships between firms and directors, coding a tie between them if a given director sits on a firm's corporate board (see Figure 1.4). By taking the projection of this bipartite network, it was possible to construct a network of relationships between firms. Using this method, a relationship between firms

FIGURE 1.4
Taking the Projection of the Director Interlock Network

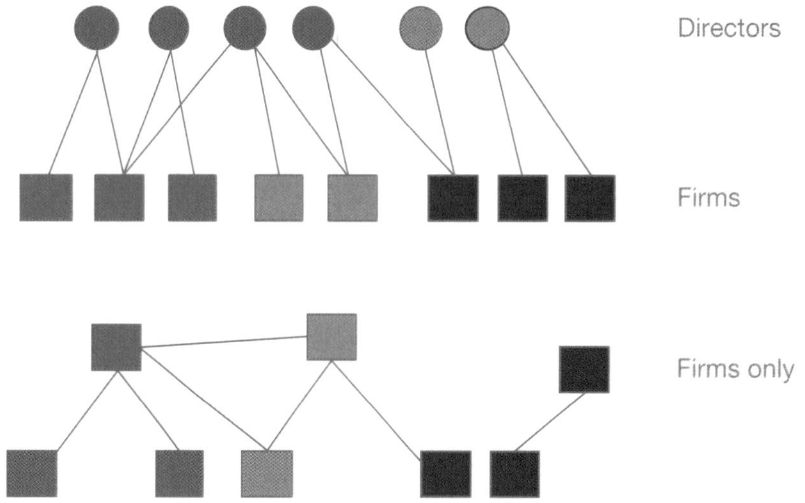

NOTE: Colors and shading are intended to make the figure more legible and do not represent specific information.

was said to exist if the same director sat on the boards of both companies. This is a network implementation that captures the phenomenon of board interlock. It is also possible to take a projection of the relationships between directors, in which there is a tie if they sat on the same corporate board. For the present study, we have limited our analysis to the network of firms, though it is also possible to study the linkages between people who both sit on a given corporate board.

Shareholder data were reported in the Osiris database with the firms or owners identified by unique codes and including a corresponding share of ownership. Shareholder networks were constructed by using these data to code directional ties between companies where the weight of the network edge was the percentage of shareholder ownership. Data on ownership by non-Asian companies and by individual investors were not included in this analysis.

Within our dataset, firms are associated with the country in which they are incorporated, rather than by the exchange on which they are listed. Thus, there are Chinese firms that are listed on the Hong Kong or New York Stock Exchange that are still identified as Chinese companies.

This analytical approach is necessarily limited in that it captures only public relationships between firms. Many relationships between firms are purposely hidden, and many others are coordinated through mechanisms other than director interlock (Dieleman, 2007). Even with the limitations of this study, the results described below provide an unprecedented view of interfirm relationships across the entire region.

CHAPTER TWO

Results and Analysis

Director Interlock Networks

The network of board of director interlocks provides a useful entry point through which to examine the evolving character of the Asian economy. Network diagrams showing the full dataset of firms from the years 2006 and 2020 are depicted in Figure 2.1. These visualizations clearly convey several important trends in Asian development: the growing number of public firms, the increasing size of the Chinese economy, and the dense interlocking of directorates among a core set of firms in the region.

Individual firms are represented in Figure 2.1 as colored circles with the color representing the country in which the firm is incorporated. Director interlocks between firms are represented by gray lines connecting them. The algorithm used to generate these network diagrams spaces the firms as far away from one another as possible, as though they were repelling each other, while the ties between firms pull them closer together.

The result is that firms linked to one another are positioned in proximity to each other, and when many relationships are present between a set of firms, the result is a tight cluster of firms. The most densely

FIGURE 2.1
Comparison of Board Interlock Networks in Asia, 2006 and 2020

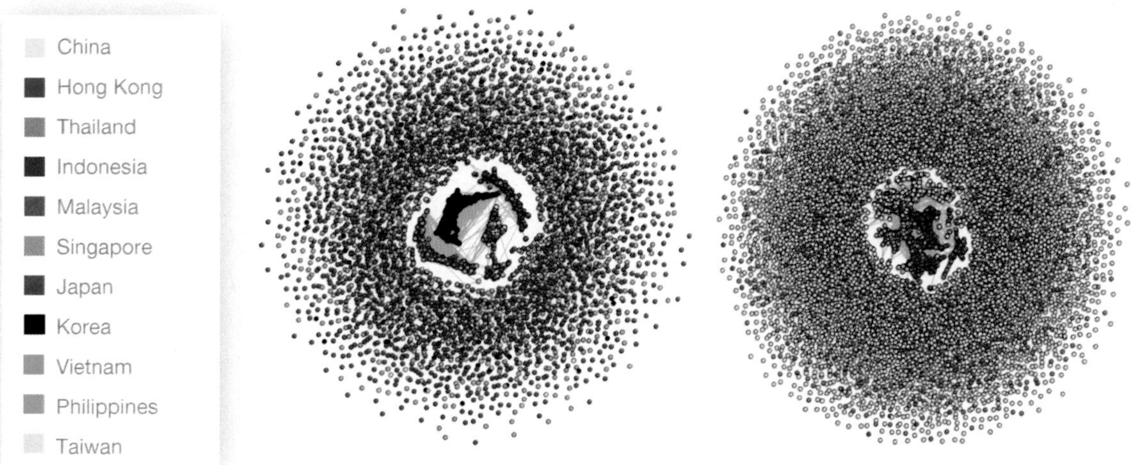

2006

Firms:	9,111
Directors:	50,625
Relationships:	30,481

2020

Firms:	21,552
Directors:	363,929
Relationships:	54,594

connected sets of firms are depicted at the center of the diagram, with smaller clusters and eventually isolated firms positioned at correspondingly greater distances from the center. The significance and structure of these clusters will be described and analyzed in the following sections.

In both images the basic structure remains the same. The bulk of the companies, represented by the disconnected circles on the outside of the image, are not connected to any other firms. This reflects the fact that the board members of most companies do not sit on any other corporate boards. Rather, they draw their directors from friends, confidants, or managers from private companies.

The most obvious transformation between the two periods is the growth of the overall number of listed companies between 2006 and 2020. This growth is evident in the increased number of circles in the later image, while the change in the proportions of the different nationalities of the firms in the sample is evident in the dramatic increase in the number of Chinese (yellow) firms. The shifting composition of the sample is evident from a tabular summary of the number of firms from each country in each year's dataset (see Table 2.1). The dramatic increase in the number of cross-border ties—those that connect firms incorporated in different countries—from 243 to 2624 speaks to the rapid internationalization that occurred during the period under study, particularly because the growth in the number of international ties (9.8 × increase) far outpaced the increase in the number of firms in the dataset (1.4 × increase).

TABLE 2.1

Number of Exchange-Listed Firms, 2006 and 2020

Country	2006	2020	Change (percentage)
China	1,451	10,192	602
Hong Kong	251	208	–17
Indonesia	282	546	94
Japan	3,239	3,768	16
Laos	0	5	N/A
Malaysia	995	772	–22
Myanmar	0	2	N/A
Philippines	201	171	–15
Korea	1,516	2,106	39
Singapore	482	529	10
Taiwan	268	1,670	523
Thailand	404	682	69
Vietnam	22	901	3,995
TOTAL	9,111	21,552	137
Total firms	9,111	21,552	137
Total ties	30,481	54,594	79
Domestic ties	30,238	51,970	72
Cross-border ties	243	2,624	980

NOTE: Bureau van Dijk data; data includes only firms with more than $10M annual revenue.

Increasing Integration of Asian Economies

Significantly, both networks contain a central core of tightly connected firms that are closely linked to one another through directors with shared board memberships. Focusing on the central core of firms that are connected to at least one other company through a shared board member, we clearly see the relative lack of integration within the regional economy in 2006 relative to 2020 (see Figures 2.2 and 2.3). Because this network-visualization format minimizes the total lengths of all interfirm ties depicted on the graph, strongly interconnected sets of firms, such as those in China or Korea, are bundled closely together, while more weakly interconnected groups, such as the firms of Malaysia, are more spread out. In some cases, sets of firms within a country are related through shared director positions, but the firms that share these ties do not connect with any other outside firms. In this case, the firms appear as disconnected clusters, with the dots representing the component firms stacked on top one another. As a result, while 18 Indonesian firms were connected with other firms through director interlocks, they appear as isolated clusters around the edge of the image in Figure 2.2.

One striking feature of this structure is the insularity of these economies. Within this sample of Asian firms, there were only 243 cases where firms from different countries shared the same director. Korean firms, for example, had no interlocks with foreign firms, and they were rare between other countries. One reason for these national divides is the strong linguistic and cultural differences between the countries in the sample. In most countries, board meetings would typically be held in the local language, which made it quite difficult for a foreign member to participate.

A second significant feature of these networks is the high percentage of firms that share directors with other companies (see Figure 2.2). In China, Malaysia, Korea, and Hong Kong, more than 75 percent of the

FIGURE 2.2
Asian Firms Connected by Board of Director Interlocks, 2006

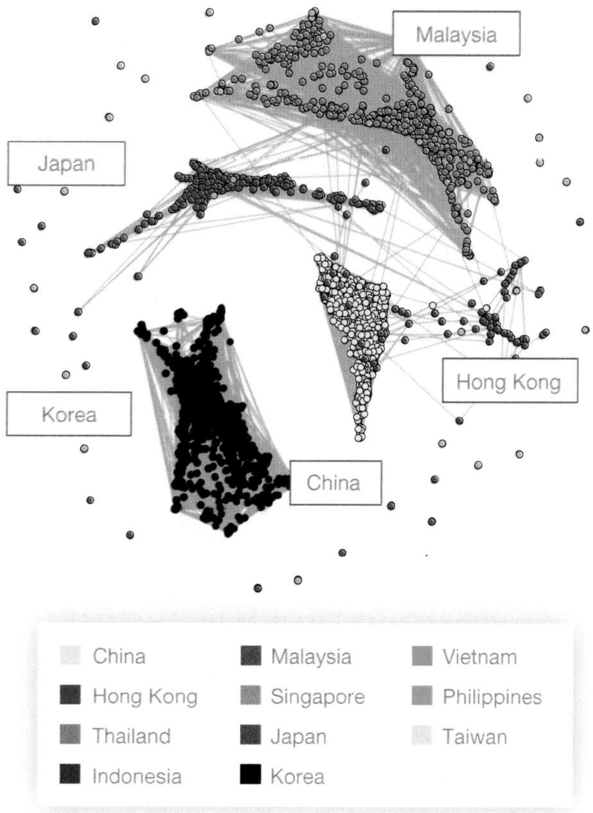

	Total firms	Networked firms	Percentage networked
China	1,451	1,422	98
Hong Kong	251	188	75
Indonesia	282	58	21
Japan	3,239	1,199	37
Korea	1,516	1,448	96
Malaysia	995	975	98
Philippines	201	76	38
Singapore	482	106	22
Taiwan	268	22	8
Thailand	404	93	23
Vietnam	22	0	0
TOTAL	9,111	5,587	61.3

FIGURE 2.3

Asian Firms Connected by Board of Director Interlocks, 2020

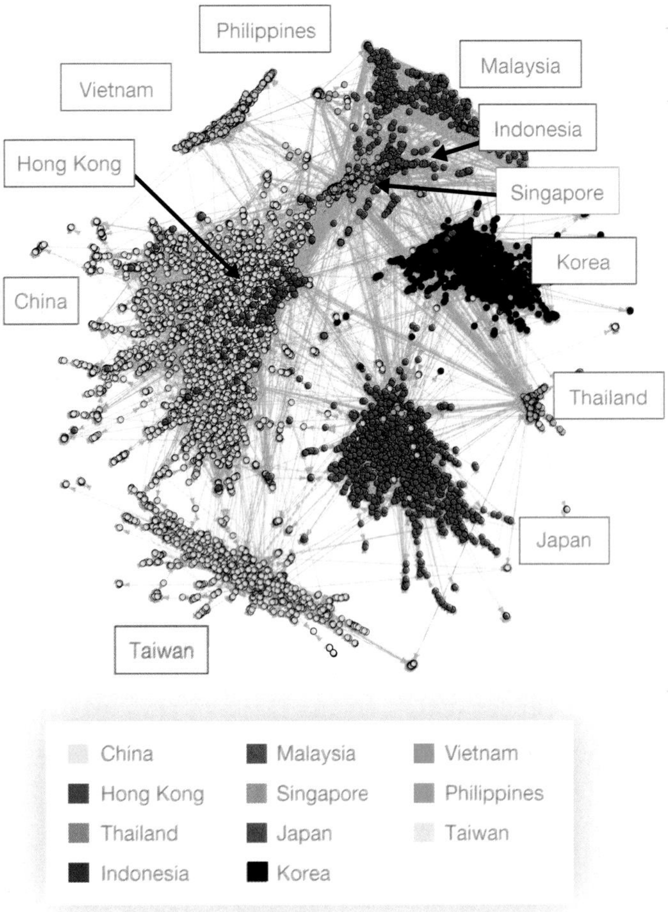

	Total firms	Networked firms	Percentage networked
China	10,192	4,221	41.4
Hong Kong	208	201	96.6
Indonesia	546	484	88.6
Japan	3,768	3,217	85.4
Korea	2,106	1,944	92.3
Laos	5	1	20.0
Malaysia	772	763	98.8
Myanmar	2	1	50.0
Philippines	171	168	98.2
Singapore	529	525	99.2
Taiwan	1,670	1,237	74.1
Thailand	682	645	94.6
Vietnam	901	804	89.2
TOTAL	21,552	14,211	65.9

firms in the country shared a director with another firm.[1] This is likely due to the normative inclusion on corporate boards in these countries of high-profile individuals, who linked the firms together. The potential importance of shared directorates in coordinating actions within the economy is thus highlighted. Additionally, different norms for public offerings meant that it was more common for subsidiaries in countries such as Malaysia to be listed as independent companies than was the case in Indonesia and other nations with less developed equity markets.

By 2020, the interlock structure of the region had evolved considerably. The total number of firms in the sample more than doubled, from 9,111 to 21,552, while the number of cross-border director ties had increased more than tenfold, from 243 to 2624. Thus, while for the most part clusters of firms from the same nation are well defined, indicating that it is more common to share directors within rather than between countries, there is significantly more interaction between national groups. Additionally, the Korean, Chinese, and Japanese business networks have become more dispersed, meaning that while firms still share directors, they draw from a larger pool of candidates.

As shown in Figure 2.3, nine of the 13 countries in the dataset have more than 80 percent of their firms incorporated into this network of shared directors, as do roughly two-thirds of all listed firms in the region.

---

1 As noted above, this value is likely inflated due to the name-matching strategy employed to identify shared directorships.

Other economies, like that of Taiwan, have become dramatically more networked over time. Moreover, as the visual representation of the network makes clear, Hong Kong is now essentially embedded within the network of Chinese firms, while in 2006 it retained its relative autonomy.

The Strength of Interfirm Ties: Multiple Relationships Between Firms

To address the potential concern that many board of director ties are inconsequential and have no instrumental value in sharing information or coordinating corporate activities, we also modeled the network of firms that share two or more directors—a situation that indicates that there is a partnership or at least a close relationship between the firms in question. While a single shared director could simply be a well-respected expert who sits on two firms boards simply by coincidence, firms that share two directors are more likely to be members of the same corporate group or to have their policies coordinated with one another.

These multidirector interlocks likely indicate the presence of a stronger and perhaps qualitatively different relationship than do single shared directors. While sharing a single director may be relatively inconsequential or allow indirect influence through information sharing, multiple shared directors suggests that the relationship is intended to promote something more concrete, such as improved corporate coordination or direct monitoring of a business partner.

Figure 2.4 shows the frequency of these more intensive interlocks and provides a visualization of their network structure.

FIGURE 2.4
Network of Firms Sharing Two or More Directors, 2020

Country	Within country	To other Asian countries
China	690	95
Hong Kong	107	113
Indonesia	277	81
Japan	747	46
Korea	2,142	14
Malaysia	1,702	84
Philippines	366	39
Singapore	572	148
Taiwan	509	17
Thailand	522	56
Vietnam	405	7

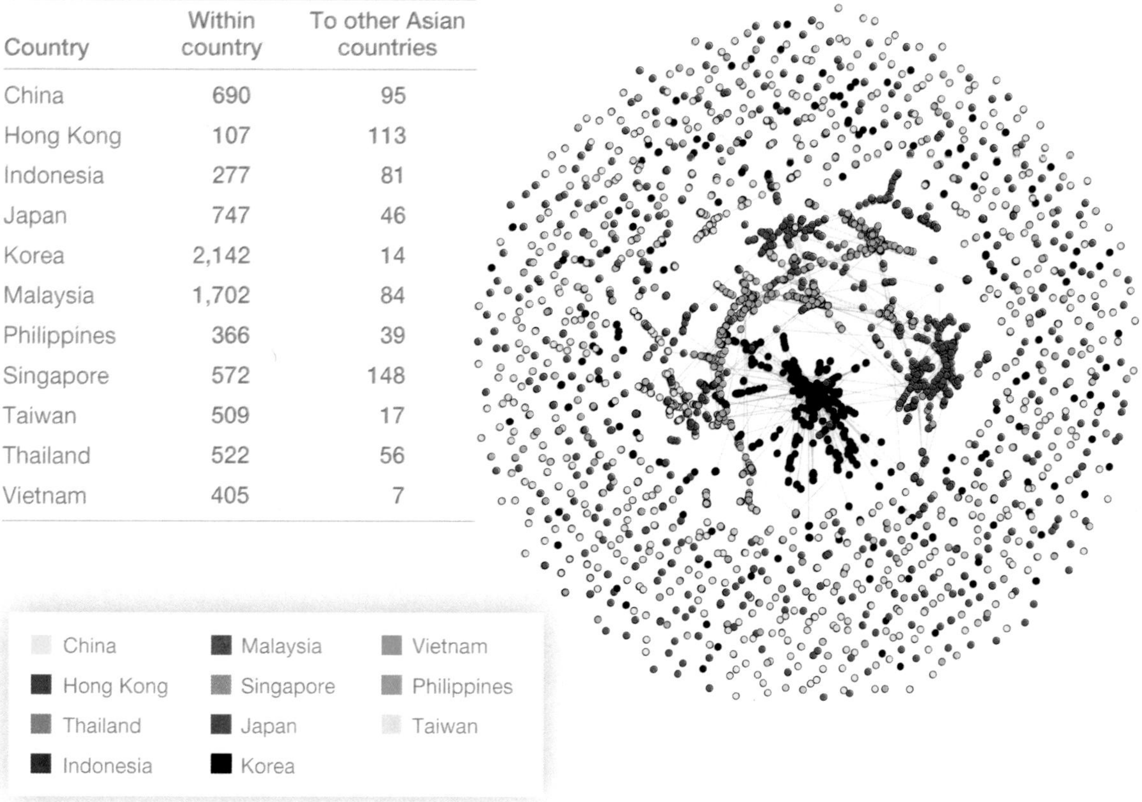

It is notable that a significant number of multidirector ties are present both within and between countries. If we interpret these multidirector interlocks as an indication of the purposeful formation of an interfirm relationship with some instrumental function, the significant number of cross-border ties suggests that these interlocks provide firms with a mechanism to promote coordination or exert influence across international borders.

Shareholding Networks

Networks of equity relationships were also analyzed. Table 2.2 presents data on equity ownership in Asian firms by shareholders in Asia. These shares can be held by other firms (see below) but are also held by privately held institutions, public investment accounts such as Chinese state-owned Assets Supervision and Administration Commission, or individuals.

One concern with this broad view of equity holdings is that many equity stakes are too small to confer any significant degree of control to the holder. For example, a shareholder's owning 0.5 percent of a firm's shares may represent a significant investment in the target company, but because the shareholder controls only 0.5 percent of the voting rights, it may not be able to use the shares to influence firm policymaking.

By filtering on a more restrictive criterion and looking only at shareholders that own 10 percent or more of a target company, we find the pattern of controlling equity stakes becoming more legible. As

TABLE 2.2
Shareholding Relationships Within Asia, 2006 and 2020

	2006		2020		Change (percentage)	
Country	Within Country	To Other Asian Countries	Within Country	To Other Asian Countries	Within Country	To Other Asian Countries
China	7,196	11	142,088	244	1,875	2,118
Hong Kong	47	106	561	2,233	1,094	2,007
Indonesia	15	0	1,941	23	12,840	N/A
Japan	5,073	135	49,257	1,133	871	739
Korea	824	2	7,760	180	842	8,900
Laos	0	0	9	1	N/A	N/A
Malaysia	651	6	13,004	266	1,898	4,333
Myanmar	0	0	6	0	N/A	N/A
Philippines	15	0	1,647	17	10,880	N/A
Singapore	59	177	5,941	1,686	9,969	853
Taiwan	190	36	3,803	187	1,902	419
Thailand	44	2	8,361	57	18,902	2,750
Vietnam	0	0	4,422	0	N/A	N/A

Table 2.3 shows, there are fewer of these larger equity stakes, and their prevalence has been diminishing over time.[2] While the number of large cross-border equity holdings has increased in some cases, the data suggests that these are not common control structures due to their expense and the potential for negative publicity that large foreign investments might attract.

An alternative view of the data analyzes only interfirm relationships, excluding equity ownerships by individuals, private firms, or state-owned entities. Data from 2020, presented in Table 2.4, reveal that large equity stakes are relatively uncommon as mechanisms for cross-border partnership. They are more frequent within the larger and more established economies of Japan, Taiwan, and Korea. By comparison, board of director interlocks are far more frequent than shareholding relationships in every country but Japan.

TABLE 2.3

Significant Shareholding Relationships Within Asia, 2006 and 2020

	2006		2020			
	Shareholding Connections of More Than 10-Percent Equity		Shareholding Connections of More Than 10-Percent Equity		Change (percentage)	
Country	Within Country	To Other Asian Countries	Within Country	To Other Asian Countries	Within Country	To Other Asian Countries
China	3,114	6	156	38	–95	533
Hong Kong	24	63	44	58	83	–8
Indonesia	11	0	11	6	0	N/A
Japan	2,280	65	326	64	–86	–2
Korea	599	0	609	3	2	N/A
Laos	0	0	0	0	0	N/A
Malaysia	96	3	717	22	647	633
Myanmar	0	0	0	0	0	N/A
Philippines	11	0	35	1	218	N/A
Singapore	46	39	158	78	243	100
Taiwan	186	10	412	11	122	10
Thailand	38	1	36	8	–5	700
Vietnam	0	0	20	0	0	N/A
TOTAL	6,405	187	2,524	289	–61	55

NOTE: Data on ownership percentages are missing from many records, so the data presented here likely understate the true frequency of large equity stakes. The 2006 data represent direct ownership, while the 2020 data record the ultimate owner.

[2] The percent ownership is missing in 16 percent of records, so this may understate the true number of large equity stakes in the sample. Second, the 2006 data records direct ownership, while the 2020 data records the ultimate owner. This is significant because, for example, some shares in Hong Kong were listed as held by Hong Kong Securities Clearing Company Nominees Limited, which holds shares to act as a common nominee in order to facilitate trades on the stock exchange.

TABLE 2.4
Comparison of Governance Relationships Among Firms, 2020

	Shareholding Connections		Shareholding Connections of More Than 10-Percent Equity		Board of Director Interlocks Between Firms	
Country	Within Country	To Other Asian Countries	Within Country	To Other Asian Countries	Within Country	To Other Asian Countries
China	4,378	61	52	16	8,808	1,169
Hong Kong	24	170	13	16	458	1,009
Indonesia	102	4	3	2	1,288	297
Japan	16,493	891	220	56	8,972	247
Korea	1,755	54	150	3	10,442	133
Laos	1	0	0	0	0	4
Malaysia	708	64	19	5	8,090	630
Myanmar	0	0	0	0	0	1
Philippines	83	3	4	0	1,150	145
Singapore	89	341	20	25	4,347	1,123
Taiwan	1,158	71	224	6	2,515	123
Thailand	637	13	19	3	2,759	270
Vietnam	421	0	4	0	3,123	90
Total	25,849	1,672	728	132	51,952	5,241[a]

NOTES: *Equity holdings* refers to the ultimate owner. Data on ownership percentages are missing from many records, so these data likely understate the true frequency of large equity stakes.

[a] Board interlocks between countries are counted twice, as though they originated once in each country. This doubles the total number of ties, but provides useful data on the countries in which cross-border relationships are most common.

Regional Patterns of Interlock

One significant feature of business networks in Asia is the difference between the interlock networks of East Asian (China, Taiwan, Japan, Korea, Hong Kong) and Southeast Asian (Thailand, Malaysia, Singapore, Indonesia, Philippines, Vietnam, Laos, Myanmar) nations (see Figure 2.5).

As these diagrams show, director interlock is significantly more common in Southeast Asia than it is in East Asian economies; this is apparent from the large number of black, cross-border ties that represent the extensive integration of the boards of companies in the region. The data on the frequency of these cross-boundary director interlocks are presented in Table 2.5.

Though East Asian economies have three times as many listed firms as the Southeast Asian region, they have only one and a half times as many shared directors. This unbalanced ratio is driven, in part, by the larger numbers of directors holding multiple board seats in Southeast Asia (see Table 2.6). Recent studies of interlocking directorates in the American economy have documented a dramatic reduction in directors serving on multiple boards (Chu and Davis, 2016). While this has erased the traditional "inner circle" of directors that coordinated the American economy throughout the twentieth century, it appears that "super directors" with numerous board seats continue to play an important function in the economies of Asia.

FIGURE 2.5

Board of Director Ties in East Asia and Southeast Asia, 2020

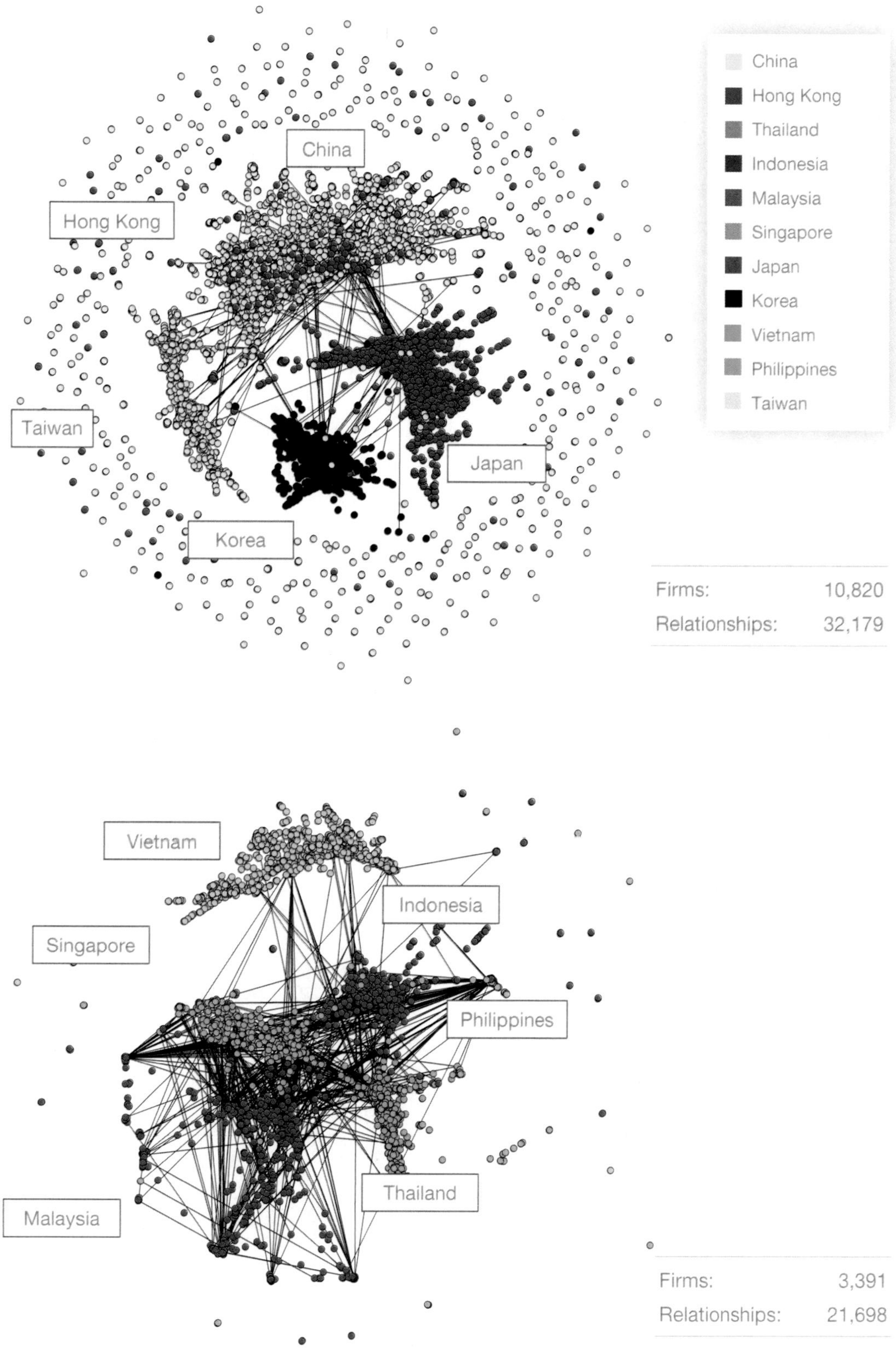

NOTE: Only relationships between firms in different countries are shown. They are represented as black lines.

TABLE 2.5

Director Interlocks Among Asian Countries, 2020

Country	China	Hong Kong	Indonesia	Japan	Korea	Malaysia	Philippines	Singapore	Thailand	Taiwan	Vietnam
China	8,808	774	6	27	30	30	2	245	7	43	5
Hong Kong	774	458	4	40	3	23	16	123	12	18	0
Indonesia	6	4	1,288	22	17	67	26	155	31	2	6
Japan	27	40	22	8,972	37	17	9	31	38	9	17
Korea	30	3	17	37	10,442	12	8	8	4	3	11
Malaysia	30	23	67	17	12	8,090	27	403	35	10	6
Philippines	2	16	26	9	8	27	1,150	33	12	8	4
Singapore	245	123	155	31	8	403	33	4,347	92	12	20
Thailand	7	12	31	38	4	35	12	92	2,759	16	19
Taiwan	43	18	2	9	3	10	8	12	16	2,515	2
Vietnam	5	0	6	17	11	6	4	20	19	2	3,123

NOTE: Because director interlocks are not directional, the rows and columns can be read in either direction. The table is symmetrical around the diagonal axis.

TABLE 2.6

Number of Board Seats Held by Directors in East and Southeast Asia, 2020

Number of Board Seats Held	Count of Directors of East Asian Firms	Count of Directors of Southeast Asian Firms	Percentage of Directors of East Asian Firms	Percentage of Directors of Southeast Asian Firms
1	284,387	48,158	92	86
2	18,433	5,386	6	10
3	3,712	1,515	1	3
4	1,044	635	0.3	1
5	309	291	0.1	1
6	138	117	<0.1	0.2
7	39	75	<0.1	0.1
8	21	44	<0.1	0.1
9	9	11	<0.1	<0.1
10	9	12	<0.1	<0.1
11	2	12	<0.1	<0.1
12	2	11	<0.1	<0.1
13	2	2	<0.1	<0.1
14	4	5	<0.1	<0.1
15+	3	22	<0.1	<0.1

NOTE: If directors have board seats in both regions, they are counted twice—once in each column.

The Role of the Ethnic Chinese Diaspora in Interfirm Networks

One factor promoting the denser international networks in Southeast Asia is the presence of an economically prosperous ethnically Chinese population whose shared linguistic and ethnic ties provide an important basis for cross-border integration in the region. In addition, their presence provides a mechanism that allows Chinese firms to partner with firms in Southeast Asian countries more easily than their other East Asian counterparts. While this analysis can offer only a suggestive account of the role of Chinese entrepreneurs in shaping the interfirm networks in the region, their potentially significant role should not be overlooked.

Chinese capitalists and entrepreneurs have historically held significant economic positions throughout Asia. Chinese merchants had for centuries acted as traders and business intermediaries throughout Southeast Asia, and in the 1930s, Chinese businessmen still played important roles in the economies of the region (Reid, 1988). During World War II, indigenous Chinese businessmen often suffered at the hands of the Japanese. After the war, the livelihoods of successful businessmen were further disrupted when the colonial administrations that they had served and European businesses that they had been connected with were dismantled and nationalized during the independence movements. In Indonesia, for example, the Chinese businessmen who had assisted the Dutch regime found themselves marginalized after independence because their Dutch education and privileged position within the colonial regime became worthless as soon as the colonial system was dismantled.

However, even as the fortunes of the previous generation of Chinese businessmen diminished, few local entrepreneurs moved to take their places. In Malaysia, Indonesia, and Thailand, for example, social roles had become well entrenched, and indigenous peoples preferred to farm or enter the government than to enter business, which they viewed as a low-prestige profession (Geertz, 1976; Polsiri and Wiwattanakantang, 2007). This vacuum was quickly filled by recent immigrants from mainland China, who often had access to capital and transregional kinship networks, which gave them advantages over local businessmen. Liem Sioe Liong, the founder of Indonesia's Salim Group and one of Indonesia's richest men, was a native of the Fujian province, for example. While he fostered the myth that he had arrived penniless, interviews with his son reveal that he had relied on help from his Chinese contacts to become an established businessman even before he met his protector and benefactor, the future president Suharto (Dieleman, 2007).

The concentration of wealth in the hands of this ethnic minority helped to foster a sense of shared identity and cemented the position of Chinese expatriates as the dominant entrepreneurs and capitalists of the region. Ethnic Chinese were subjected to periodic waves of xenophobia and ethnic backlashes and accordingly sought the protection of well-placed political and military elites to protect their assets. Moreover, the threat of ethnic strife gave them an added incentive to buy the protection of elites through bribes, advantageous business contracts, and corporate directorships, which in turn gave them added advantages and preferential access to resources, contracts, and business concessions. Xenophobia and cultural resentment of the Chinese had the effect of preventing Chinese businesspeople from forming the basis of a politically autonomous middle class, as they might have under other circumstances. Instead, they were fundamentally dependent on the government for protection, as well as for favors, and were unable to unite politically with indigenous business owners who often viewed them as ethnic competitors. As a result of these processes, big business in postwar Southeast Asia was often controlled by Chinese capitalists who were natural partners for businesses in mainland China seeking to expand to new markets (Robinson, 1986). When the ties between China and these overseas Chinese were described in the 1990s, the so-called bamboo network of ethnic Chinese businesspeople and entrepreneurs throughout Asia was seen as a resource that could provide valuable capital, know-how, and connections to help build China's economy (Weidenbaum and Hughes, 1996).

This network of overseas Chinese capitalists has been a significant factor in regional development. Seventy-five percent of the population of Singapore is ethnically Chinese, and many of its residents maintain strong personal and historical ties with China. As other scholars have noted, Singapore is an "oasis" within

Southeast Asia, an island of Chinese ethnicity surrounded by Muslim states (Kunio, 1988, p. 118). The large population of ethnic Chinese businesspeople in Southeast Asia is a resource that Chinese firms can draw on when entering a new region (Rae and Witzel, 2008; Suryadinata, 2017; Weidenbaum and Hughes; 1996).

Today, China is actively courting these ethnic Chinese populations. As Xi Jinping observed in a 2016 speech, "There are tens of millions of Chinese overseas compatriots (*haiwai qiaobao*), all of whom are members of one big Chinese family. . . . They are making important contributions to the development of the Chinese nation and . . . fostering the close and friendly cooperation between the Chinese people and people of other countries" (quoted in Suryadinata, 2017, p. 20). In addition to a centralized effort by the Chinese government to leverage foreign ethnic Chinese populations (Suryadinata, 2017), relationships based on these shared ethnic ties also occur organically. In the view of several businesspeople interviewed for this study, Chinese firms often seek out local partners who have some ties to China.[3] In some cases, these relationships may seem relatively tenuous, such as having ancestors who participated in the Chinese Communist Revolution or who have family ties to the same province as Chinese corporate leaders, but the shared ethnic roots or historical ties confer a certain legitimacy to foreign partners. In the view of those interviewed, this desire for partners with some association with China is particularly pronounced because China has such a large domestic market. Thus, many firms are able to focus on domestic expansion and do not have to cultivate external markets and partners to the degree common among firms in smaller nations. The result is that they do not have extensive networks of contacts in many markets abroad and instead seek out partners that have some externally legible markers of legitimacy and preexisting ties to China. This logic might also explain the reason why Taiwanese firms do not appear to share this preference for partners of shared ethnicity. With a smaller internal market and with a longer history of integration with partners throughout the region, they have not chosen to forge as many partnerships as firms from mainland China.

Viewed through this historical lens, one factor driving cross-border directorships of Southeast Asian firms may be the reliance on external social relationships to the ethnic Chinese communities throughout the region. The importance of this diaspora for cross-border relationships is made apparent by the inclusion of firms from China and Hong Kong in the network map of Southeast Asia (see Figure 2.6). Viewed in this context, Singapore's role as hub for firms with ties to both Chinese and other Southeast Asian firms is clearly evident. By building on the historical network of ethnic ties within the region, Chinese firms are able to forge partnerships through board representation to a degree that other nations such as Japan, Korea, or even Australia and the United States are unable to emulate. The question of whether they will be able to exert regional influence through these channels will be considered in the following section.

Understanding the Implications of China's Network Position

What would it mean if increasing numbers of Chinese firms had directors on the boards of Singaporean or Malaysian firms? If we assume that board members can, at times, play an active role in coordinating the actions of the firms that they advise and promoting communication and allowing monitoring, if not collusion between them, increased board interlock would help synchronize the actions of the nations' economies (Mizruchi, 1996; Pfeffer and Salancik, 1978).

Among business theorists who believe that board interlock is a powerful tool of corporate governance, these networks are thought to "facilitate a community of interest among the elite of the corporate world that supplants the competitive and socially divisive ethos of an earlier stage of capitalism with an ethic of cooperation and a sense of shared values and goals" (Bowman, 1996, p. 21). Even if we adopt the more measured view that

[3] Interviews were conducted in person by the author in July 2018 in Los Angeles, Hong Kong, and Singapore. Interviews were not for attribution, and the interviewees were promised anonymity to ensure that responses would be as candid as possible.

FIGURE 2.6
Director Interlock Among Asian Firms with Only Cross-Border Ties Shown, 2020

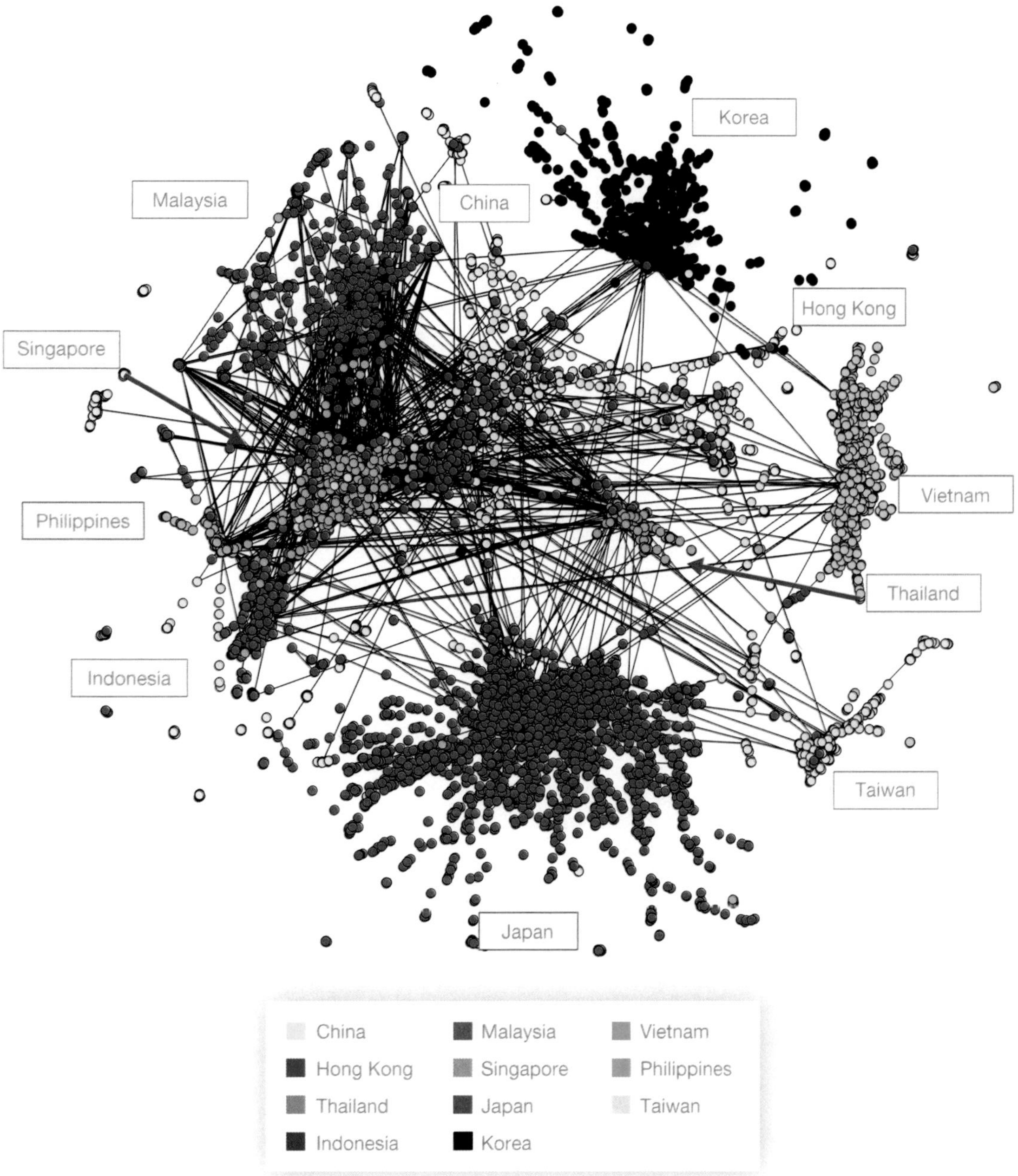

director interlocks are "mundane but consequential" (Davis, 1996, p. 154), the disproportionate representation of Chinese firms on boards in the region will likely have long-term positive impacts on Chinese integration into the regional economies, encouraging path-dependent decisions such as the adoption of norms, including, for example, Chinese technology standards, which could produce a durable orientation toward China.

If we accept these networks of relationships as important, a related question is how broad the sphere of influence exerted by foreign directors might be. Just as director networks enabled the spread of behaviors such as the "poison pill" that protects against hostile takeovers to be shared among firms and that spawned coordinated adoption of the multidivisional corporate structure, influence can be spread through indirect ties to reach more distant parts of the network (Davis, 1991; Palmer, Jennings, and Zhou, 1993). In Asia, interlocking directorates are a modern structure that has, to some degree, been overlaid onto an older, more relationship-based model for business. In his analysis of interfirm relationships among firms owned by ethnic Chinese businesspeople in Singapore, the sociologist Xiaowei Zang (1999, p. 864) argues that interlocking directorates are "a modern form of maintaining personal relationships in the Chinese business community." Viewed through this lens, interlocking directorates are instrumental in maintaining "*guanxi* capitalism," (*guanxi* is a Chinese term meaning "connectedness"), in which business is conducted through networks based on relationships built on mutual trust and obligation and reinforced by actions such as information sharing with business partners (Zang, 1999).

Viewing director interlock networks as a potential conduit for communication and coordination suggests a way to visualize the question of how far the influence of these networks extends. One might imagine that direct interlock ties between a Chinese company and a foreign firm indicate that the firms coordinate their activities or collaborate in significant ways. The question is the degree to which firms indirectly connected to Chinese firms—that is, sharing directors with a firm that is connected to a Chinese firm—are included within the Chinese-directed sphere of influence. Using the director network data, we can identify the firms that are indirectly connected to a Chinese firm because they share a director with a firm that has a director who also serves on the board of a Chinese company (see Figure 2.7). Viewed from a social exchange standpoint, these have a director who will meet the Chinese firm's director at the board meetings of a third company. One can imagine how these shared meetings could result in information exchange and collaboration, whether planned or inadvertent. As the graphic represents, this networking strategy encompasses most companies in Hong Kong and Singapore as well as a substantial number of firms in Malaysia and Japan. If,

FIGURE 2.7
Firms with Direct or Indirect Ties to Chinese Firms

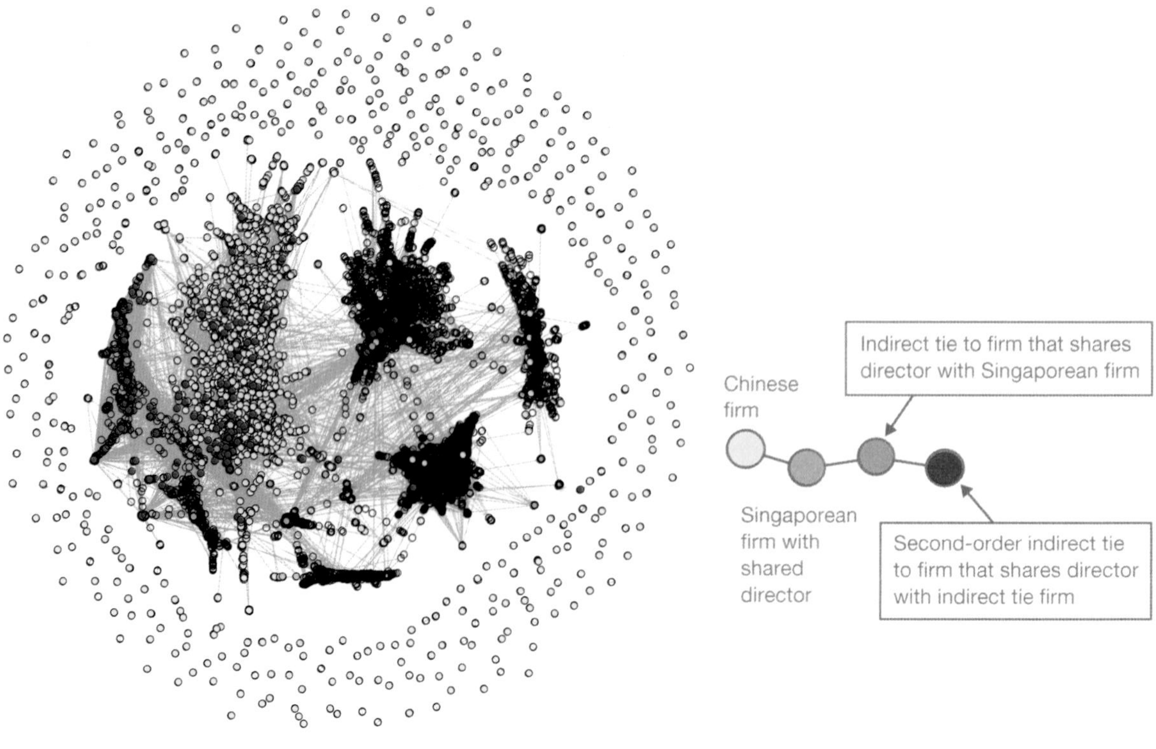

NOTE: Chinese firms and all firms connected to them directly or indirectly appear as colored circles in this graph. Firms that are not part of this network are depicted as empty circles.

as it seems at the time of this writing, companies incorporated in Hong Kong become less independent and uniformly adopt Chinese policies, the reach of these networks would be even greater still.

The reach of these direct and indirect networks has increased dramatically in recent years. Comparing the growing number of firms with direct or indirect interlocking directorate ties to Chinese firms, we can see that these types of interlock relationships are becoming more significant over time (Tables 2.7 and 2.8).

It is not clear from these analyses whether the broadening reach of these networks is the result of a conscious strategy or simply a side effect of the increasing integration of Asian business relationships. In either case, however, the potential influence of these director relationship networks should be carefully evaluated over the coming years.

TABLE 2.7

Number of Firms with a Director Who Serves on the Board of a Chinese Firm, 2018 and 2020

Country	2018	2020	Change (percentage)
Hong Kong	134	148	10
Indonesia	3	5	67
Japan	10	17	70
Korea	5	10	100
Malaysia	8	19	138
Philippines	1	2	100
Singapore	69	94	36
Taiwan	11	34	209
Thailand	1	4	300
Vietnam	0	2	N/A

TABLE 2.8

Number of Firms with a Director Who Has an Indirect Tie to a Chinese Firm, 2018 and 2020

Country	2018	2020	Change (percentage)
Hong Kong	246	188	–24
Indonesia	22	65	195
Japan	68	183	169
Korea	20	403	1,915
Malaysia	106	348	228
Philippines	10	25	150
Singapore	399	455	14
Taiwan	40	125	213
Thailand	30	74	147
Vietnam	0	27	N/A

The Role of Domestic Networks as Moderators or Magnifiers of Foreign Influence

One remaining question is whether the different network structures within Asian economies serve to stifle or facilitate the spread of foreign influence. Rather than imagining the developing economies of the region as filled with small companies that are at risk of takeover from their larger, expanding neighbors, this analysis has shown that it is more accurate to think of them as enmeshed in a larger, network ecosystem in which the majority of firms are already aligned.

The structure and extent of these national networks are increasingly visible from data on director interlocks. In 2006, interlocks were still relatively rare, meaning that these networks were not yet visible as formalized director networks in many of the economies of the region, though networks of obligation and coordination existed in other forms. Today, however, the widespread adoption of Western-style governance norms has resulted in the use of directors networks as a mechanism of interfirm governance. As a result, by 2020, all of the Asian nations analyzed, with the exception of China, had a single large network component that included 60 percent or more of a nation's firms (see Table 2.9). This networking of the majority of a nation's firms into a larger connected network is a striking testament to the degree of informal coordination that is possible within these nations. In countries such as Malaysia, nearly every public firm is connected in a single-director overlap network, meaning that monitoring, influence, and the spread of normative practices extend throughout the economy.

The business groups that dominate many of the economies of the region are natural matrices for the formation of these networks. Because business groups are marked by coordinated action and shared governance, the members of these groups typically place members on each other's boards to increase information sharing and trust and to foster coordination. In nations with weak formal institutions, these groups are thought to internalize the functions typically provided through public means; the groups provide internal mechanisms for dispute resolution, provide access to financing, and lower the risk of default. Moreover, their

TABLE 2.9

Comparison of Director Interlock Networks in Asian Countries, 2020

Country	Number of Firms	Number of Interfirm Relationships	Number of Connected Firms in Main Network Component	Number of Cross-Border Firms in Main Connected Component	Percentage of Firms in Main Component	Percentage of Connected Firms with International Ties
China	10,192	8,808	3,439	607	33.7	17.7
Hong Kong	208	458	157	136	75.5	86.6
Indonesia	546	1,288	447	140	81.9	31.3
Japan	3,768	8,972	3,056	176	81.1	5.8
Korea	2,106	10,442	1,934	93	91.8	4.8
Malaysia	772	8,090	761	284	98.6	37.3
Philippines	171	1,150	167	57	97.7	34.1
Singapore	529	4,347	520	312	98.3	60.0
Taiwan	1,670	2,515	1,099	78	65.8	7.1
Thailand	682	2,759	632	113	92.7	17.9
Vietnam	901	3,123	801	39	88.9	4.9

economic importance often develops in parallel to strong political ties. In Indonesia, for example, 70 percent of exchange-listed firms are affiliated with one of the business groups, and the head of one the largest, the Bakrie Group, was also the head of a major political party. The result is that there is an extensive and increasingly formalized set of networks in place that tie the firms of national economies together even as the number of ties between countries increases.

The degree of a nation's interaction with other firms in other nations can be calculated as the percentage of the firms in this connected network that have connections to firms in other nations. This metric reveals a substantial divide among the countries of Asia. The density of these cross-border connections speaks to the internationalization of the region. The majority of firms in Hong Kong and Singapore share ties with other nations, while about one-third of the firms in Indonesia, the Philippines, and Malaysia have cross-border ties. By contrast, Taiwan, Japan, Korea, and Vietnam have only a single-digit percentage of external ties.

One important question is the degree to which the existence of extensive interlock networks within each of these national economies might moderate or magnify the influence of foreign firms. One perspective is that the existence of dense networks of domestic ties serves to dampen the impact of foreign directors. While a foreign board member might be able to influence the decisions of a single company to coordinate with a foreign firm, the existence of numerous other connections to domestic companies could well dilute this influence. The nature and extent of these competing domestic ties can be assessed by calculating different measures of network topology.

Several network measures can be used to gauge the degree of integration provided by these networks (see Table 2.10). The mean path metric measures the average distance between firms in the network. Here, it refers to the number of shared directors that separate one firm from another. In the case of tightly networked nations such as the Philippines, this means that on average, firms are separated by only 2.4 directors. A second measure, average degree, is the average number of other companies that a given firm is connected to, with higher values indicating greater degrees of integration. The variation among nations is particularly striking, with companies in Singapore, the Philippines, Malaysia, and Korea connected to more than ten other

TABLE 2.10

Comparison of Network Characteristics of Director Interlock Networks in Asian Countries, 2020

Country	Diameter of Main Component	Mean Path in Main Component	Mean Degree in Main Component	Transitivity in Main Component
China	21	5.7	4.88	0.28
Hong Kong	13	3.7	5.62	0.31
Indonesia	18	4.4	5.68	0.31
Japan	20	4.9	5.81	0.14
Korea	20	3.4	10.79	0.12
Malaysia	9	3.0	21.26	0.65
Philippines	12	2.4	13.77	0.32
Singapore	9	2.8	16.72	0.36
Taiwan	24	5.9	4.43	0.34
Thailand	21	3.6	8.72	0.24
Vietnam	15	3.9	7.80	0.36

firms and firms in Taiwan and China connected to fewer than five. A final measure, transitivity (also known as the clustering coefficient), assesses the density of ties within the network by measuring the overall probability for adjacent nodes to be interconnected within the network. Japan and Korea have relatively low values in this measure, likely because the pyramidal business groups in both countries economize on direct relationships with apex companies by delegating these functions to lower tiers of the pyramid. In Malaysia, by contrast, dense interlocked webs are the norm.

While Tables 2.9 and 2.10 demonstrate the dramatic differences in network topology between countries, the impact of these variations in network structure and the degree to which they might blunt or enhance foreign influence is unclear. One possibility is that tight networks of domestic ties might exert a conservative influence and serve to minimize any influence that foreign directors could exert. In this view, in nations such as Malaysia and the Philippines that have significant numbers of firms with foreign ties, their domestic firms are tightly interlocked with one another, creating a particularly dense web of relationships that may blunt the influence of foreign directors.

A second effect is that the dense network of relationships might also serve as conduits for information exchange, which might aid foreign efforts to monitor or influence a domestic economy. From this perspective, while dense networks of firms might impede efforts to influence a specific firm, it is often the companies that are most central to the network that have foreign relationships. To assess the question of whether the companies with international ties are more centrally positioned in a nation's director network, the betweenness centrality scores of companies with and without international board interlocks were compared with a Welch two-sample *t*-test (see Table 2.11).[4]

TABLE 2.11

Testing Whether Firms with International Director Ties Have Higher Betweenness Than Other Firms in the Country: Results of Welch Two-Sample *t*-Test

Country	Degrees of Freedom	*p*-Value
China	666.58	<0.001**
Hong Kong	280.19	0.347
Indonesia	219.81	0.003**
Japan	178.99	<0.001**
Korea	92.968	0.081
Malaysia	412.68	0.001**
Philippines	77.322	0.108
Singapore	686.46	0.550
Thailand	134.25	0.001**
Taiwan	98.364	0.331
Vietnam	40.59	0.010**

NOTE: ** indicates *p*-values of less than 0.01.

[4] *Betweenness centrality* is a measure of the relative centrality of the position of a node in a network. Its calculation is based on the number of paths between other nodes that go through a given node, with higher values indicating a more central position.

In many Southeast Asian countries, the international ties are connected to the boards most central in the network. In Hong Kong and Singapore, which are each distinguished by their large percentages of firms with international board ties (see Table 2.9), there is not a significant difference between the two groups of firms. Likewise, there is no difference between the network positions of the two groups of firms in counties such as Korea and Taiwan that have few international ties, though it is unclear why this is also true for Philippine firms.

The implication is that foreign directors on these boards are particularly well positioned to gather information and exert influence. High-betweenness firms are strategically positioned within the network and are defined as being on a large number of the shortest paths between other nodes. The implication is that information, norms, or behaviors that spread through these nodes are therefore able to quickly spread through the network. Likewise, their central position renders the directors of these firms well positioned to monitor other firms in the national economy. The alternative is that these firms are so deeply enmeshed in the networks of their domestic economies that they are able to resist foreign influence. Further research will be required to assess the specific mechanisms at work in any given national context.

CHAPTER THREE

Conclusions

This report has examined and mapped the interfirm relationships that extend throughout the Asia-Pacific region. By doing so, it provides for the first time an integrated regional assessment of the network structure of interfirm governance in Asia. Its descriptive analysis highlights the existence and importance of these interfirm networks and is organized around three principle aims: (1) to identify the extent of shared governance ties between Asian firms, both within and between countries; (2) to look at the regional differences in the extent and structure of these ties; and (3) to consider the implications of these ties for China's evolving role in the regional economy, both in terms of the structure of the networks connecting China to other countries and the potential role of ethnic Chinese populations in the region in facilitating the formation of these networks.

Interfirm Networks in Asia

This report argues that interfirm networks are particularly significant within Asia. Historically, many of the region's economies have been dominated by business groups or conglomerates that, as one researcher described, stand out as "islands of planned coordination in a sea of market relations" (Richardson, 1972, p. 883). The members of these groups, of which the Chinese *qiye jituan*, the Korean *chaebol*, and the Japanese *keiretsu* are familiar examples, act in concert to support shared group-level strategies. In contrast to the vision of competitive and atomized firm decisionmaking championed by American economic theorists, these multifirm networks are central to the operation of the regional economy.

These conglomerates are particularly significant because they were often controlled by ethnic Chinese families, whose family and shared ethnic ties extended across the region. The concept of this so-called bamboo network was popularized in the 1990s and identified as a driving force behind China's economic growth (Weidenbaum and Hughes, 1996). As researchers at the time noted, a relatively small group of foreign-born ethnic Chinese business tycoons dominated the business landscape in Asia and preferred to conduct their business dealings within their extended families or deal with other ethnic Chinese businesspeople with whom they shared some connection.

However, the shock of the Asian Economic Crisis of the late 1990s and the internationalization of business over the following decades has transformed the postwar system of family and relationship based business relationships. In desperate need of external capital due to the economic crisis and with weakening patronage networks resulting from the ensuing political turmoil, many firms shifted away from their closely held family structures and adopted a more standardized model in order to better access international credit markets. Foreign financiers required that firms become increasingly "credible" and "transparent" in the eyes of potential creditors, and Asian firms largely adopted management practices and systems of financial control that were prescribed by the gatekeepers of the global financial system (Yeung, 2006).

Despite these changes, the role of family, kinship, and ethnic networks in business networks has not disappeared (Luo and Chung, 2005). While firms have adopted the norms of international business, including the shareholders, boards of directors, and professional managers that comprise the three central pillars of modern corporate governance (Monks and Minnow, 2012), they have often adopted these structures in form only (Young et al., 2008). Thus, while many aspects of business operation have modernized to align with global norms, they have not removed the traditional reliance on family and ethnic ties. Rather, these traditional models have, in many cases, co-opted these new formal business strictures to create networks that allow business groups to coordinate the actions member firms both within and across national borders (Colpan, Hikino, and Lincoln, 2018; Yeung, 2006).

While existing research on Asian business groups has documented the important roles played by these interfirm networks in particular national contexts, little work has been done to map the ways these groups extend beyond national borders or to understand the ways that they influence the creation of cross-border ties that include them. This report uses data made available as a result of the standardization of business structure to develop a more systematic view of the ways in which governance relationships span national boundaries to create regional webs of influence, coordination, and control.

The report has focused on two types of interfirm relationships that represent two of the formal governance methods employed by public companies—the voting rights of shareholders and the board of directors.

The first of these, equity ownership, is straightforward. Companies, banks, other organizations, or individuals can buy stock in another company both to share in its profits and to exercise control through shareholder voting. Anything more than a 50-percent equity stake effectively confers control by guaranteeing majority voting rights to a shareholder, though minority stakes also provide the ability to exert influence over firm decisionmaking. While our analysis shows that equity holdings are used in Asia, particularly within the economies of Japan, Korea, and Taiwan, they are not common due to the high cost of acquiring a significant equity stake. The analysis showed that in 2020, there were 728 equity holdings of 10 percent or more between firms of the same country, and 132 that crossed national borders. While a significant interfirm governance strategy that provides the potential for formal control rights, they were much less common than the second type of relationship we analyzed.

Board interlock, in which one company invites representatives of a second firm to sit on its board of directors, forms a second formalized type of interfirm relationship. This type of relationship was much more common, with 51,952 interfirm ties within countries, and 5,241 between them. Board members are predominantly executives from other firms who typically serve on the boards of one or two other firms (Chu and Davis, 2016; Davis, 1996). These board memberships bring executives into contact with other members with comparable ties, creating a broad and complex set of linkages with other firms (Boyd and Hoskisson, 2010). As other researchers have described, this network of relationships "facilitate[s] a community of interest among the elite of the corporate world that supplants the competitive and socially divisive ethos of an earlier stage of capitalism with an ethic of cooperation and a sense of shared values and goals" (Bowman, 1996, p. 21) and represents a "trans-corporate network, overarching all sectors of business" (Useem, 1986, p. 53). More recently, researchers have taken interlocking directorates as a proxy for the existence of a formal link between two companies and have used them to map the structure of business networks (Colli and Vasta, 2018).

In Asia, interlocking directorates are a modern structure that has, to some degree, been overlaid onto an older, more relationship-based model for business. In his analysis of interfirm relationships among firms owned by ethnic Chinese businesspeople in Singapore, the sociologist Xiaowei Zang (1999, p. 864) argues that interlocking directorates are "a modern form of maintaining personal relationships in the Chinese business community." Viewed through this lens, interlocking directorates are instrumental in maintaining "*guanxi* capitalism" in which business is conducted through networks based on relationships built on mutual trust and obligation and are reinforced by actions such as information sharing with business partners (Zang, 1999).

The Structure and Increasing Internationalization of Interfirm Networks

The network analysis presented in this report demonstrates that the structure and extent of these national and international business networks is increasingly visible from data on director interlocks. In 2006, interlocks were still relatively rare, meaning that these networks were not yet visible as formalized director networks in many of the economies of the region, though networks of obligation and coordination existed in other forms. Today, however, the widespread adoption of Western-style governance norms has resulted in the use of director networks as a mechanism of interfirm governance. As a result, by 2020, all of the Asian nations analyzed, with the exception of China, had a single large network component that includes 60 percent or more of a nation's firms (see Table 2.9). This networking of the majority of a nation's firms into a large connected network is a striking testament to the degree of informal coordination that is possible within these nations. In countries such as Malaysia, nearly every public firm is connected in a single-director overlap network, meaning that monitoring, influence, and the spread of normative practices extend throughout the economy.

The past 15 years have seen a dramatic increase in the number of cross-border director interlock ties—those that connect firms incorporated in different countries—from 243 in 2006 to 2,624 in 2020. This speaks to the dramatic internationalization that occurred during the period of this study, particularly because the growth in the number of international ties (9.8 × increase) far outpaces the increase in the number of firms in the dataset (1.3 × increase).

The significance of these networks is confirmed by mapping the structure of multidirector relationships in addition to the network of single shared director ties. While a single shared director could simply be a well-respected expert who sits on two firms' boards simply by coincidence, firms that share two directors are more likely to be members of the same corporate group or to have their policies coordinated with one another. These multidirector interlocks likely indicate the presence of a stronger and perhaps qualitatively different relationship than do single shared directors. While sharing a single director may be relatively inconsequential or allow indirect influence through information sharing, multiple shared directors suggests that the relationship is intended to promote something more concrete, such as improved corporate coordination or direct monitoring of and influence over a business partner.

The analysis of these networks demonstrates that even when using a more stringent definition of what constitutes an interfirm relationship, and recording them only when firms share two or more directors, there are a significant number of interfirm ties within Asia. In 2020, there were 8,039 multidirector relationships between firms in the same country and 700 cross-border ties in which firms in different countries shared two or more directors.

The Role of China in Regional Interfirm Networks

What would it mean if increasing numbers of Chinese firms had directors on the boards of Singaporean or Malaysian firms? If we assume that board members can, at times, play an active role in coordinating the actions of the firms that they advise, promoting communication, and allowing monitoring, if not collusion, increased board interlock would help synchronize the actions of the nations' economies (Mizruchi, 1996; Pfeffer and Salancik, 1978).

Viewing director interlock networks as a potential conduit for communication and coordination provides a means to understand the question of how far the influence of these networks extends. One might imagine that direct interlock ties between a Chinese company and a foreign firm indicate that the firms coordinate their activities or collaborate in significant ways. The question is the degree to which firms indirectly connected to Chinese firms—that is, sharing directors with a firm that is connected to a Chinese firm—are included within the Chinese-directed sphere of influence. Using the director network data, we can identify the firms that are indirectly connected to a Chinese firm because they share a director with a firm that has a director who also serves on the board of a Chinese company (see Figure 2.7). Viewed from a social exchange standpoint, these firms have a director who will meet the Chinese firm's director at the board meetings of a third company. One can imagine how these shared meetings could result in information exchange and collaboration, whether planned or inadvertent.

Mapping these indirect networks reveals that most countries in Asia now have a hundred or more firms with indirect ties to Chinese firms. Furthermore the number of these relationships has increased significantly since 2018. Additional analysis demonstrates that in most countries, foreign director ties are concentrated among firms with the higher betweenness centrality—that is, they connect foreign firms with corporate boards that are particularly well positioned to gather information and exert influence. High-betweenness firms are strategically positioned within the network and are defined as being on a large number of the shortest paths between other nodes. The implication is that the information, norms, or behaviors that spread through these nodes are able to quickly spread through the network. Likewise, their central position renders the directors of these firms well positioned to monitor other firms in the national economy. The other alternative is that these firms are so deeply enmeshed in the networks of their domestic economies that they are able to resist foreign influence because they are tied to so many other firms that can moderate foreign influence.

Looking Forward

The growing number of cross-border governance relationships between Asian firms highlights the importance of understanding interfirm relationships from a policy perspective. Thus, rather than viewing national economies as composed of individual firms that exist under the umbrella of a national economy, this analysis highlights the importance of better understanding the role played by these networks in regional integration and their function of advancing or retarding high-level initiatives intended to foster particular patterns of partnership and collaboration.

References

Borsuk, R., and Chng, N. (2014). *Lim Sioe Liong's Salim Group: The Business Pillar of Suharto's Indonesia.* ISEAS Publishing.

Bowman, S. R. (1996). *The Modern Corporation and American Political Thought: Law, Power, and Ideology.* Pennsylvania State University Press.

Boyd, B. K., and Hoskisson, R. E. (2010). "Corporate Governance in Business Groups," in A. M. Colpan, T. Hikino, and J. R. Lincoln (eds.), *The Oxford Handbook of Business Groups.* Oxford University Press, 670–695.

Burt, R. S. (1992). *Structural Holes: The Social Structure of Competition.* Harvard University Press.

Chang, S. J. (2003). "Ownership Structure, Expropriation, and Performance of Group-Affiliated Companies in Korea," *Academy of Management Journal,* 46:238–253.

Chu, J., and Davis, G. (2016). "Who Killed the Inner Circle? The Decline of the American Corporate Interlock Network," *American Journal of Sociology,* 122:714–754.

Colli, A., and Vasta, M. (2018). "Italy: Enduring Logic and Pervasive Diffusion," in A. Colpan and T. Hikino (eds.), *Business Groups in the West: The Evolutionary Dynamics of Big Business.* Oxford University Press, 273–307.

Colpan, A., and Hikino, T. (2010). "Foundations of Business Groups: Towards an Integrated Framework," in A. M. Colpan, T. Hikino, and J. R. Lincoln (eds.), *The Oxford Handbook of Business Groups.* Oxford University Press, 15–66.

Colpan, A., Hikino, T., and Lincoln, J. R., eds. (2018). *Business Groups in the West: The Evolutionary Dynamics of Big Business.* Oxford University Press.

Csardi, G., and Nepusz, T. (2006). "The Igraph Software Package for Complex Network Research," *InterJournal,* Complex Systems, no. 1695.

Davies, W., and Marlow, I. (2019). "China Seeks to Buy Control of Hong Kong Companies, Report Says," Bloomberg, September 12.

Davis, G. F. (1991). "Agents Without Principles? The Spread of the Poison Pill Through the Intercorporate Network," *Administrative Science Quarterly,* 36:583–613.

______. (1996). "The Significance of Board Interlocks for Corporate Governance," *Corporate Governance,* 4:154–159.

Dieleman, M. (2007). *The Rhythm of Strategy: A Corporate Biography of the Salim Group of Indonesia.* Amsterdam University Press, 2007.

Geertz, C. (1976). *The Religion of Java.* University of Chicago Press.

Granovetter, M. (1995) "Coase Revisited: Business Groups in the Modern Economy," *Industrial and Corporate Change,* 4:93–130.

Granovetter, M. S. (1985). "Economic Action and Social Structure: The Problem of Embeddedness," *American Journal of Sociology,* 91:481–510.

______. (2005). "Business Groups and Social Organization," in N. Smelser and R. Swedberg (eds.), *Handbook of Economic Sociology,* 2nd ed. Princeton University Press.

Katz, R. (1998). *Japan, the System That Soured: The Rise and Fall of the Japanese Economic Miracle.* M. E. Sharpe.

Khanna, T., and Rivkin, J. W. (2006). "Interorganizational Ties and Business Group Boundaries: Evidence from an Emerging Economy," *Organization Science,* 17:333–352.

Kranton, R. (1996). "Reciprocal Exchange: A Self-Sustaining System," *American Economic Review,* 86: 830–851.

Kunio, Y. (1988). *The Rise of Ersatz Capitalism in Southeast Asia.* Oxford University Press.

Lee, K., and Kang, Y.-S. (2010). "Business Groups in China," in A. M. Colpan, T. Hikino, and J. R. Lincoln (eds.), *The Oxford Handbook of Business Groups.* Oxford University Press, 210–236.

Lin, B., Chase, M. S., Blank, J., Cooper III, C. A., Grossman, G., Harold, S. W., Moroney, J. D. P., Morris, L. J., Ma, L., Orner, P., Shih, A., and Kim, S. (2020). *Regional Responses to U.S.-China Competition in the Indo-Pacific: Study Overview and Conclusions*, Santa Monica, Calif.: RAND Corporation, RR-4412-AF. As of July 21, 2021: https://www.rand.org/pubs/research_reports/RR4412.html

Lincoln, J., and Gerlach, M. (2004). *Japan's Network Economy: Structure, Persistence, and Change.* Cambridge University Press.

Lincoln, J., Gerlach, M., and Ahmadjian, C. (1996). "Keiretsu Networks and Corporate Performance in Japan," *American Sociological Review*, 61:67–88.

Lincoln, J. R., and Sargent, M. (2018). "Business Groups as Networks," in A. Colpan and T. Hikino (eds.), *Business Groups in the West: The Evolutionary Dynamics of Big Business.* Oxford University Press, 95–122.

Lockwood, W. W. (1968). *Economic Development of Japan: Growth and Structural Change.* Princeton University Press.

Luo, X., and Chung C. N. (2005). "Keeping It All in the Family: The Role of Particularistic Relationships in Business Group Performance During Institutional Transition," *Administrative Science Quarterly*, 50:404–439.

Mizruchi, M. S. (1996). "What Do Interlocks Do? An Analysis, Critique, and Assessment of Research on Interlocking Directorates," *Annual Review of Sociology*, 22:271–298.

Monks, R. A. G., and Minnow, N. (2012). *Corporate Governance*, 5th ed. John Wiley & Sons.

Palmer, D. P., Jennings, D., and Zhou, X. (1993). "Late Adoption of the Multi-Divisional Form by Large U.S. Corporations: Institutional, Political and Economic Accounts." *Administrative Science Quarterly*, 38:100–131.

Pfeffer, J., and Salancik, G. R. (1978). *The External Control of Organizations: A Resource-Dependence Perspective.* Harper and Row.

Piore, M., and Sabel, C. (1984). *The Second Industrial Divide: Possibilities for Prosperity.* Basic Books.

Polsiri, P., and Wiwattanakantang, Y. (2007). "Thai Business Groups: Crisis and Restructuring," in Sea-Jin Chang (ed.), *Business Groups in East Asia.* Oxford University Press, 147–178.

Porter, M. E. (1996). Competitive Advantage, Agglomeration Economies, and Regional Policy," *International Regional Science Review*, 19.

Rae, I., and Witzel, M. (2008). *The Overseas Chinese of South East Asia: History, Culture, Business.* Palgrave Macmillan.

Reid, A. (1988). *Southeast Asia in the Age of Commerce, 1450–1680.* Yale University Press.

Richardson, G. B. (1972). "The Organisation of Industry," *The Economic Journal*, 82: 883–896.

Robinson, R. (1986). *The Rise of Capital.* Allen and Unwin.

Saxenian, A. (1994). *Regional Advantage: Culture and Competition in Silicon Valley and Route 128.* Harvard University Press.

Suryadinata, L. (2017). *The Rise of China and the Chinese Overseas: A Study of Beijing's Changing Policy in Southeast Asia and Beyond.* ISEAS Publishing.

Useem, M. (1986). *The Inner Circle: Large Corporations and the Rise of Business Political Activity in the U.S. and U.K.* Oxford University Press.

Vedres, B., and Stark, D. (2010). "Structural Folds: Generative Disruption in Overlapping Groups," *American Journal of Sociology*, 115:1150–1190.

Watts, D. J. (1999). *Small Worlds: The Dynamics of Networks Between Order and Randomness.* Princeton University Press.

Weidenbaum, M. (1998). "The Bamboo Network: Asia's Family-Run Conglomerates," *World View*, 10.

Weidenbaum, M., and Hughes, S. (1996). *The Bamboo Network: How Expatriate Chinese Entrepreneurs Are Creating an Economic Superpower in Asia.* The Free Press.

Williamson, O. (1975). *Markets and Hierarchies*. The Free Press.

______. (1985). *The Economic Institutions of Capitalism*. The Free Press.

World Bank Data. (2020). "GDP (Constant 2010 US$)," web database. As of October 7, 2020: https://data.worldbank.org/indicator/NY.GDP.MKTP.KD

Yeung, H. W. C. (2006). "Change and Continuity in Southeast Asia Ethnic Chinese Business," *Asia Pacific Journal of Management*, 23:229–254.

Yiu, D. W., Lu, Y., Bruton, G. D., and Hoskisson, R. E. (2007). "Business Groups: An Integrated Model to Focus Future Research," *Journal of Management Studies*, 44:1551–1579.

Young, M. N., Peng, M. W., Ahlstrom, D., Bruton, G. D., and Jiang, Y. (2008). "Corporate Governance in Emerging Economies: A Review of the Principal-Principal Perspective," *Journal of Management Studies*, 45:196–220.

Zang, X. (1999). "Research Note: Personalism and Corporate Networks in Singapore," *Organization Studies*, 20:861–877.